FRANK SPRINGER AND NEW MEXICO

Frank Springer, c. 1915. Courtesy of Museum of New Mexico, Neg. 7886.

Frank Springer and New Mexico

From the Colfax County War to the Emergence of Modern Santa Fe

DAVID L. CAFFEY

Texas A&M University Press
College Station

Second printing, 2007

The paper used in this book meets the minimum requirements of the American National Standard for Permanence of Paper for Printed Library materials, Z39.48-1984.
Binding materials have been chosen for durability.

∞

LIBRARY OF CONGRESS CATALOGING IN PUBLICATION DATA

Caffey, David L., 1947–
Frank Springer and New Mexico : from the Colfax County War to the emergence of modern Santa Fe/ David L. Caffey.—1st ed.
p. cm.
Includes bibliographical references (p.) and index.
ISBN 1-58544-464-2 (cloth : alk paper)
1. Springer, Frank, 1845–1927. 2. New Mexico—Biography. 3. Frontier and pioneer life—New Mexico. 4. Maxwell Land Grant Company—History. 5. Lawyers—New Mexico—Biography. 6. Journalists—New Mexico—Biography. 7. Santa Fe (N.M.)—History—19th century. 8. Violence—New Mexico—Colfax County—History—19th century. 9. Colfax County (N.M.)—History—19th century. 10. Paleontologists—United States—Biography. I. Title.
F801.S68C34 2006
978.9'2204'092—dc22

ISBN-13: 978-1-60344-004-2 (pbk.)
ISBN-10: 1-60344-004-6 (pbk.)

For Linda Mitchell Davis,
and to the Memory of J. Leslie Davis

Contents

Illustrations

Preface

When Frank Springer jounced over Raton Pass and into the territory of New Mexico on a spring wagon one winter day in 1873, he beheld the facing skyline of an estate that would command much of his effort and attention over the next fifty-four years. Just twenty-four years of age, Springer was recently arrived from Iowa, where he had left a promising law practice for what he hoped would be an even greater opportunity in the Southwest.

From the wagon track that led across the plain to Cimarron—a segment of the "Mountain Branch" of the Santa Fe Trail—Springer could see the distant shapes of Agua Fria Peak, Black Mountain, and Clear Creek Mountain. As he neared the village of Cimarron he could look westward to the bare summit of Baldy—the object of much excitement on the part of miners and prospectors. Before these soaring masses were the lower tablelands; Rayado Mesa and Urraca Mesa; the steady decline of Tooth of Time Ridge, and the piñon-juniper foothills. Farther north the horizon revealed the pointed top of Costilla Peak; still farther north, in Colorado, the summit of Culebra Peak rose more than fourteen thousand feet above the sea.

The country surveyed from these heights was one of immense beauty and abundant resources—grass and timber, wild game, precious metals, and a vast bed of commercial-grade coal. It was also a stage upon which dramatic and sometimes violent events played out during Springer's ten-year residence in Cimarron.

In the course of twelve summers as a seasonal employee at the Philmont Scout Ranch roughly a century later, I covered a good portion of the same country, once known as the Beaubien and Miranda or Maxwell Land Grant. I explored what remained of the gold mines, mills, and mining camps on the slopes of Baldy Mountain and tramped out the abandoned, overgrown, and much eroded grades of the Cimarron and Northwestern and St. Louis, Rocky Mountain and Pacific railroads—lines that existed to carry coal, coke, and timber out of the mountains.

Under summer stars I heard and told the traditional stories of "Uncle Dick" Wootton and his toll road over Raton Pass; of Clay Allison, who in a fit of pique, possibly heightened by consumption of alcohol, threw the press and type of the local paper in the Cimarron River one long-ago night; of the iconoclastic preacher O. P. McMains, who sought to avenge the death of a fellow clergyman and stayed more than two decades to fight the Maxwell Land Grant Company on behalf of settlers who contended that the land on which they had located was public domain. This was the stuff of campfire tales.

I had been to the town of Springer, but not for some years did I learn about the man for whom it was named and of his long association with the Maxwell Land Grant Company, his defense of the company's title to the property before the U.S. Supreme Court, and his involvement in the development of railroads, coal mines, ranching, and the building of the Eagle Nest Dam. Still later I learned of Frank Springer's long association with Edgar Lee Hewett in the founding of the New Mexico Normal University, and then of the School of American Archaeology and the Museum of New Mexico. To discover that this same man had carried on a parallel career as a paleontologist, amassing collections, conducting research, and publishing his findings in the presses of leading scientific institutions, was somewhat astounding.

Among his associates, Springer was sometimes referred to as a "many-sided man"—a man of varied interests who had accomplished much in diverse and demanding fields. He was that, to be sure, but there were other complexities in his makeup as well. He was at once an introvert who enjoyed nothing more than an evening with his fossils, and a very public man who led in the development of major commercial enterprises, mustered support for cultural and educational institutions, held elective office as a member of the territorial council, and regularly delivered public orations to large audiences. He placed his faith in science and quietly but firmly eschewed organized religion, yet was held in high regard in communities and societies in which churchgoing was the norm among respected citizens. Outwardly modest, he nonetheless took pleasure in publishing legal arguments, court opinions, speeches, scientific papers, and other documents that reflected favorably on his achievements, and sending them to friends and colleagues.

Frank Springer was part of the migration of well-educated, non-Hispanic white men who came to the Southwest following ratification of the Treaty of Guadalupe Hidalgo and attained economic and political influence in the new territories. He saw New Mexico as an undeveloped land on the verge of rapid economic growth, soon to be fueled by the arrival of the railroad, and he was eager to avail himself of opportunities presented for financial success. He was an ambitious young man, and in those early times he was not overly concerned with the social consequences of his ambition. Neither was he a bully or a land grabber, and he did not personally engage in the sort of aggressive land grant speculation commonly associated with the "Santa Fe Ring." His chief client, the Maxwell Land Grant and Railway Company, was engaged in such speculation by virtue of its purchase of the claim originally granted to Carlos Beaubien and Guadalupe Miranda by the Mexican government. Springer's ultimate point of reference was the law—American law—and his efforts were devoted to pursuit of the legal interests of his client.

Given the expectations of many of the westering whites and the license granted them in American law, and given the differing values and perspectives of the whites, the indigenous tribes, and the native Hispanic people, economic and cultural conflict was practically inevitable. It was part of the frontier milieu in which Springer placed himself.

When Springer arrived in Cimarron, there remained a small number of Ute and Jicarilla Apache Indians who had not moved to reservation lands designated for their use by the U.S. government. Citizens complained of occasional acts of thievery and public drunkenness among the remaining Indians and clamored for their removal; the Cimarron newspaper, edited for a time by Springer and William R. Morley, voiced the complaint in print. The government was slow to act, but within a few years Colfax County was devoid of Indians.

As attorney for the Maxwell Land Grant and Railway Company, one of Springer's first duties was to file ejectment suits against settlers, including many Hispanics and non-Hispanic whites, who did not hold legal title to the lands on which they were living, and whose claims were in conflict with those of the Maxwell Company. In the protracted struggle that followed, he was a staunch defender of the company's interest. The settlers mounted a determined resistance to the Maxwell Company's efforts to make them pay or leave, but in the end the company prevailed, due in no small part to the legal skills of Frank Springer.

Springer's credentials as a bona fide agent of Manifest Destiny are doubtful, since New Mexico had been a U.S. territory for some twenty-five years before his move to the Southwest. He may, however, be considered among those "powerful white men" who, in the view of Patricia Nelson Limerick and other contemporary historians, consolidated the gains of the conquerors and seized the upper hand economically and politically. He was also a great deal more than that.

In consequence of his successful defense of his company's title to lands surveyed under the Maxwell claim, Springer gained an economic foothold in the territory, and by virtue of his professional work and his business investments, he substantially attained the prosperity to which he had aspired. Thereafter, much of his time and resources were devoted to civic and philanthropic pursuits, and to his scientific work.

As president of the regents during the formative years of the New Mexico Normal University, Springer helped create a dynamic institution of higher education; as a territorial legislator he worked for improvements in public education. With Edgar Lee Hewett, he established cultural institutions that supported a growing community of artists and an emerging tourist trade, while preserving native antiquities and advocating their exploitation as signal assets of the region. He was influential in articulating a distinctive approach to the discipline of anthropology as it developed in the Southwest, and in determining the defining architectural style that would come to characterize Santa Fe. As president of the Maxwell Land Grant Company, he took an enterprise plagued with repeating cycles of failure and mismanagement and pressed forward with an agenda for industrial development and sound financial management. Unbeknownst to many of his professional and business colleagues, he simultaneously carried on a lifelong program of scientific investigation that produced a long list of published works and won for him the admiration of scientists in the United States and in Europe. His accomplishments in any one of these fields would have secured for him a place in history.

In an era of heightened interest in the roles of women in western American history, the slight presence of women in Springer's life story is striking. Throughout much of his life, as drawn from documentary sources, women are all but invisible. In the context of the prevailing pattern of gender roles in Springer's time, this is neither unusual nor very surprising. One might especially wish for a more revealing portrayal of Springer's wife, Josephine, and of their relationship. In the absence of much conclusive information, it can be said that they were married nearly fifty-one years; they reared seven children who led productive and responsible adult lives; and if Josephine had a serious rival for her husband's affection, it would seem to have been his work.

A passion for work, in fact, was one of the defining features of Springer's character. In reviewing Springer's life, one associate described him as "a steady and indefatigable worker," but it went beyond that. He was a compulsive worker. During the busiest years of Springer's business and professional career, he was, at times, a man seemingly driven to the limit by an overwhelming slate of tasks and obligations. There are indications that an inordinate preoccupation with his work placed a strain on Springer's family life; it certainly strained relations with his scientific collaborator, Charles Wachsmuth. Following his retirement from the practice of law, Springer continued to work, but he was working for pleasure rather than to attain financial independence, and his demeanor reflected a more relaxed and congenial person. In his last years, Springer's scientific work provided a welcome relief from the tedium of an invalid's life.

Having learned something of Frank Springer's remarkable life, I wondered aloud to western historian Lawrence R. Murphy why so little had been written about the man. Murphy, the author of a fine general history of the Cimarron country and a life of Lucien Maxwell, simply nodded and replied, "Springer badly needs a biography."

As soon as I could consider taking on the project, I wrote to Springer's grandson J. Leslie Davis, president and general manager of the CS Cattle Company at Cimarron, to express my interest in the project. Les Davis's telephone call marked the beginning of a rewarding friendship and an adventure in historical research, as I traced Springer's life through materials owned by the family and archival records located in institutions from London to Los Angeles. Springer was a highly literate person, an energetic correspondent, and a prolific producer of documents—diaries, letters, legal briefs and arguments, speeches, published articles, and scientific treatises. An abundance of primary source material was available, much of it well preserved and managed in institutional archives, and made available by helpful librarians and archivists.

I am indebted to many people and institutions whose assistance made this work possible, or made it more complete. I acknowledge with gratitude assistance received from the following institutions and organizations: Museum of Fine Arts, Santa Fe, New Mexico; Photographic Archive, Museum of New Mexico, Santa Fe; Indian Arts Research Center, School of American Research, Santa Fe; New Mexico State Records Center and Archives, Santa Fe; Philmont Museum and Seton Memorial Library, Philmont Scout Ranch, Cimarron, New Mexico; Yale University Library, New Haven, Connecticut; Peabody Museum, Harvard University, Cambridge, Massachusetts; Ernst Mayr Library, Museum of Comparative Zoology, Harvard University; Schlesinger Library, Radcliffe Institute, Harvard University; Museum of Natural History, Smithsonian Institution, Washington, D.C.; National Gallery of Art, Smithsonian Institution; National Anthropological Archives, Smithsonian Institution; Smithsonian Institution Archives; Archaeological Institute of America, Boston, Massachusetts; University of Iowa Libraries, Iowa City, Iowa; Law Library, University of Iowa, Iowa City; State Historical Society of Iowa, Iowa City; Chicago Historical Society, Chicago, Illinois; the Natural History Museum, London, England; Braun Research Library, Autry National Center/Southwest Museum, Los Angeles, California; Henry E. Huntington Library, San Marino, California; General Library, Center for Southwest Research, University of New Mexico Libraries, Albuquerque, New Mexico; Clovis Community College Library, Clovis, New Mexico; Clovis-Carver Library, Clovis; Eastern New Mexico University General Library, Portales, New Mexico; Rio Grande Historical Collections, Las Cruces, New Mexico; General Library, New Mexico Highlands University, Las Vegas, New Mexico; Citizens Committee for Historic Preservation, Las Vegas, New Mexico; Stephen H. Hart Library, Colorado State Historical Society, Denver, Colorado; Western History Collection, Denver Public Library, Denver; Arthur Johnson Memorial Library, Raton, New Mexico; Library of Congress, Washington, D.C.; U.S. National Archives and Records Administration, Washington, D.C.; and Joseph R. Skeen Library, New Mexico Institute of Mining and Technology, Socorro, New Mexico.

J. Leslie Davis, J. Wallace Davis, Gregory J. Hobbs, N. Gary Lane, Marc Simmons, and Robert R. White each reviewed parts of the manuscript and provided helpful comments and encouragement, for which I am grateful. Les Davis also made available the extensive papers and correspondence of Frank Springer in collections of the CS Cattle Company at Cimarron,

New Mexico. The materials had been carefully preserved, and in some cases annotated, by his mother, the late Ada Springer Davis, daughter of Frank Springer.

This work is respectfully dedicated to the memory of J. Leslie Davis, and to his wife, Linda Mitchell Davis. Les Davis was a proud steward of his family's legacy. It is a source of regret that his death on May 12, 2001, prevented his seeing the biography in finished form. Conversely, it is gratifying to have shared with him the pieces of the manuscript as they were produced in draft, and to have known his entire satisfaction that Frank Springer's life story was being told.

FRANK SPRINGER
AND NEW MEXICO

CHAPTER 1

A Son of Iowa

In 1846, two years before Frank Springer's birth, Iowa was admitted as the Union's twenty-ninth state. In the same year, Winnebago and Pottawatomie Indians agreed to a treaty with the U.S. government and moved to reservation lands beyond the Missouri River. These were but the latest in a series of cessions by tribes that had lived in the region prior to the coming of the whites. Writing of the years leading up to statehood, a long-time resident recalled the matter as follows: "Thus Iowa was to be relieved of an Indian population, and the whole state, with as little waste land as any other equal portion of the earth's surface, and with conditions of climate favorable to health and vigor, was opened to civilization, to the hand of industry, to the plough and the spade, to the planting of homes, to the school and the church, to representative government, and to equal laws and courts of justice."[1]

Considerable pride was expressed that Iowa had entered the Union as a "free" state, innocent of the shame of slavery, but the displacement of indigenous tribes apparently was of little concern. As settlement of the West proceeded following the Civil War, a similar litany of prospective benefits would echo time and again, enumerating the blessings that would accrue to settlers once a region could be rid of its Indian population.

Frank Springer thus grew up in a place that was imbued with the values and concerns of the frontier, where the westering impulse was well understood by white Americans and other recent immigrants. His parents had settled in Iowa before statehood and had been involved in the development of farms, communities, schools, churches, and political institutions. As a young lawyer Frank Springer's father, Francis Springer, had been advised to "go West [and] grow up with the country," and had done so, finding opportunity and even prosperity at the edge of the westward migration.

A Gentleman of the Old School

Francis Springer was born in Maine on April 15, 1811. Asked to narrate the major events of his life, he offered scant reflection on his childhood, noting only that "[c]ircumstances separated me when a lad of eleven years from my relatives."[2] Whatever the circumstances, he went to live with a farm family in Strafford County, New Hampshire. He acquired a basic education, obtained the required credential for teaching in the public grammar school, and taught two terms in rural schools, returning to his native state of Maine in 1833.

Francis Springer began to read law in the office of William Goodenow at Portland, Maine, serving intermittently as assistant editor of the *Portland Courier.* He was admitted to the bar at Portland in 1838, and soon contracted a moderate case of the "western fever." With a friend, Edward H. Thomas, he struck out for Illinois. At Cincinnati, the travelers were advised to consider settling in the rapidly developing territory of Iowa. They continued west, arriving at the capital city of Burlington, Iowa, on December 21, 1838.

Springer and Thomas spent a week acquainting themselves with influential men of the territory, calling on the governor, judges, members of the legislative assembly, and fellow members of the bar. Based on information gathered in the week's reconnaissance, they decided to locate in Louisa County and left Burlington for the nearby town of Wapello, Iowa, a place consisting of a few families and several unfinished frame houses. They became the county's first resident attorneys, handling some forty cases between them at the first term of court in Louisa County in April, 1839. In 1840 Francis Springer won election to the legislative council, a body comparable to a state senate. Upon Iowa's admission to the Union, he was elected to a four-year term in the new state senate.

In 1842 Francis Springer married Nancy Colman, daughter of Judge John M. Colman of Iowa City. Two sons died in infancy, and a daughter died at the age of two years, but Nancy gave birth to four more sons, Frank, Warren, Arthur, and Charles; and a daughter, Nellie, all born between 1848 and 1860.

Francis Springer spent much of his working life as a judge and public administrator, dividing time between his home and his work, wherever it might take him. At the conclusion of his service in the state legislature he was appointed a special agent of the federal Post Office Department, and in 1851 he was appointed register of the U.S. Land Office at Fairfield, Iowa. Returning to Louisa County, he engaged in farming and in 1854 was elected prosecuting attorney for Louisa County. Upon the death of the incumbent county judge, he succeeded to that office. He later served as district judge for the First Judicial District of Iowa, and completed his career of public service as collector of internal revenue, retiring in 1876.

Throughout his career, Francis Springer maintained an interest in politics. In 1856 he took part in a meeting called at Iowa City for the purpose of organizing the Republican Party in Iowa. He was subsequently appointed a delegate to the first national convention of the Republican Party. In the fall of 1856 he was chosen to represent Louisa County in the approaching state constitutional convention. Upon nomination by his party, he was chosen to preside over the convention. Though many times amended, the constitution crafted under his leadership has endured into the twenty-first century as the state's fundamental law.

In retirement Francis Springer lived the life of a gentleman farmer, remaining vital and active into his eighties. In 1896 the *Columbus Safeguard* paid tribute, declaring him "A Gentleman of the Old School," and relating his personal history as one of the early settlers of Louisa County.[3] Francis Springer died in 1898.

A Scholar's Life

Frank Springer was born on June 17, 1848, at Wapello, Iowa. According to a biographical sketch, he "acquired his primary schooling in a typical little log school house of those early frontier times."[4] Having exhausted the opportunities afforded in the Columbus City schools, he enrolled at the Baptist College at Burlington for the 1861–62 school term. The school offered a classical curriculum, including studies in mathematics, rhetoric, Latin, German, and music. For Springer a classical education was a practical education that had applications in law and science and led to lifelong enjoyment of varied cultural and intellectual interests.

At the age of fourteen, Springer enrolled at the State University of Iowa at Iowa City, just over thirty miles from the family home at Columbus City. He spent the next five years there and graduated with a bachelor of philosophy degree in 1867. The university had been created in 1847, but provided

Left to right: *North Hall, Old Capitol, South Hall, State University of Iowa, c. 1870. Courtesy of State Historical Society of Iowa.*

little in the way of instruction before 1860. It operated sporadically and had awarded a few degrees, but lacked a permanent home until 1857, when the capital was relocated to Des Moines and the old state capitol was turned over to the university. Springer was admitted to the "Preparatory Department" for the fall term, 1862, and entered the main university in fall of 1863. When Springer began as a student, "Old Capitol" was the only notable campus structure, but during his years in residence the institution grew rapidly. South Hall, a four-story brick building, opened in 1863, and North Hall provided added space for scientific laboratories and classrooms beginning in 1866.[5]

Springer flourished in the lively environment of the university. Even as a student, he could appreciate that he would one day miss the camaraderie of his fellows, the mentorship of his professors, and the stimulation of rigorous intellectual exercise. He also matured greatly and found a philosophical perspective in which he could believe wholeheartedly and for life.

One of the defining moments of Springer's educational experience came in the spring term of his freshman year, when the eminent naturalist Louis Agassiz stopped in Iowa City to address the students. Agassiz visited the university in March 1864, speaking on "the coral reefs of Iowa City." The topic at first astounded and puzzled Agassiz's audience, then opened up to listeners the fascinating natural history of the area in which they were living. "Prof Agassiz' lecture was a rich treat," Springer told his mother. "It was plain and at the same time profound. The simplest child could understand him, and in it he evinced a knowledge so extensive as to surprise me. He has been very highly honored by the Emperor of France, who invited him to visit & take the Professorship in the 'Academie du Paris' with any salary he wanted but he refused. He is now Prof in the Harvard University."[6]

The war between North and South was a notable distraction during Springer's first three years at the university. Daily dispatches carried news of the war and reminded students that they, too, might be called to serve. Springer knew boys who had gone to fight and he corresponded with friends who were soldiers. The fighting never came close to Iowa City, but the patriotic fervor inspired in some of the college boys a determination to enlist. In April, 1864, Springer confessed having caught the "war fever," but within the week his desire to enlist had been quashed by a stern parental veto.

By fall the Confederacy was in decline and loyal Iowans soon could rejoice in a Union victory. The surrender of the Confederate army on April 9, 1865, was cause for jubilation, but a few days later the students received a shocking report. "We have a rumor today that Mr. Lincoln was assassinated," wrote Springer on April 15, "but it can't be traced to any reliable source and I don't believe it."[7] The next day—Sunday, April 16—a horseback courier rode from Muscatine to carry news of Lincoln's death. By Springer's senior year, the fighting was over and debating societies sparred over the question of federal policy concerning reconstruction of the South.

The degree to which Springer aspired was the B.Ph., or bachelor of philosophy. Requirements were rigidly prescribed, and included courses in rhetoric, moral philosophy, zoology, physics, chemistry, geology, mathematics, political science, and French. Between lectures and recitation, laboratory exercises, gymnasium, lessons, chapel services, and participation in activities of the literary societies, he was a busy student. In his sophomore year Springer was pressed into service as the university's German language instructor.

Springer was greatly influenced by his association with Gustavus Hinrichs, assistant professor of natural philosophy and chemistry. Hinrichs arrived on the campus at about the same time Springer did and was a member of the faculty for twenty-four years, serving in the main collegiate department and later in the newly established medical department. Hinrichs rose to eminence, helping win legislative approval for a new building to house a science laboratory, but feuds with colleagues, board members, and administrators eventually proved his downfall. Of particular significance where Springer was concerned, Hinrichs was a zealous disciple of science.

The impression that Hinrichs left on Frank Springer was deep and lasting. Asked to recall the features of his experience at the university some fifty years later, Springer gave considerable attention to the influence of Hinrichs, citing, among other qualities, an infectious enthusiasm for his subject. Noted Springer, "He introduced original methods of instruction, study and laboratory work, by which students learned to observe and think, instead of committing to memory from books. Many of the methods then devised and originally introduced by him became models for teaching in schools and in text books in after years, without—I am sorry to say—Professor Hinrichs receiving the credit for them that was his due."[8]

Springer was a member of the Zetagathian Society, an oratorical club, and a frequent participant in debates and literary activities. He attended chapel services, found time for a social life, and could conveniently pursue a passion for ice-skating on nearby rivers and ponds.

The Voice of Science

In Springer's fifth and final year at the university, his outlook on the fundamental questions of life began to take definite shape. In particular, he found his views more and more in harmony with what he understood as the basic premises and methods of science. To Springer it was not simply science, the study of natural phenomena, but Science—the ultimate source of truth and knowledge, a worldview that would shape his response in matters of every sort.

A logical concomitant of Springer's approach to science as the bulwark of his philosophy was a retreat from organized religion. Such a retreat had, in fact, been in process for some time. As an impressionable youth living away from home, Springer had attended a revival and joined the Methodist church—an act that brought hearty congratulations from his grandfather Colman, but which left the boy with grave misgivings.

Troubled at what he believed to be an emotional and therefore unreliable response to Christian religion, he retreated from involvement with the church. He returned to school the following fall—1863—a more cautious and controlled being. While not prepared to forsake religion entirely, he could confide to his mother: "I am glad to hear of so many being converted at the Congregational Church. I know that I am not doing as I expected when I joined the Meth. Church, and I know I was laboring under excitement when I did it. I have not renewed my connexion with it, and never shall until I can do so of my own calm judgement. I see that the effect of a compulsory religion is not good. I know I am doing wrong, but I can't conform to the demands of religion as I thought I could. I find it hard work to do it. I will try to improve if it is possible, but shall not again join the Church soon."[9]

In his senior year, Springer's unease with religion and his newly found attraction to other methods of discovery and proof were propelling him toward a personal philosophy that was firmly rooted in science. In his final term, it all came together. Back in the routine of lecture and laboratory following the Christmas holiday, Springer devoted his final term to studies in physical and social science. He took a course in geology. As an aide for the state Geological Survey, he formed enduring friendships with William R. Morley, an engineering student, and Orestes St. John, a geologist. His commitment to science was accelerated by a series of lectures on the history of science. By February 9, Springer was ready to conclude, "The voice of Science is *truth,* always truth, in its purity and simplicity."[10]

As Springer looked toward the end of his final semester, graduation, and his imminent choice of a vocation, he found himself torn between a passion for science and more pragmatic considerations. Reflecting on his quandary, Springer told his mother, "I sometimes think I will give up everything else and settle down & devote my life to science, and the fellowship of scientific men. But the trouble is it *won't pay.* Science is so little appreciated in this country that one can hardly make a comfortable living who devotes himself to it."[11]

Springer's final oration as an undergraduate student—a valedictory address delivered at commencement exercises in June 1867—bore the title "Science as a Civilizer." Springer relished the opportunity to express himself to the large commencement audience. The finished oration developed a history of science through centuries of darkness and suppression to the dawning of scientific thought, and included a recitation of notable contributions of science to society. The speaker revealed his own conviction that "Science takes no step backward because its conclusions are founded on truth; they are themselves realities, the expression of the infinite immutability of nature herself."[12]

Frank Springer left the University of Iowa in June, 1867, with degree in hand, having developed the clarity of thought and breadth of perspective that were the promise of a university education. He took with him skills of writing, debate, analysis, and elocution that would serve him well in law and business. He had acquired a scientific background that would support his research efforts. He had forged lifelong friendships with men of ability and influence, and he was intellectually grounded in a philosophy that would lend a logical consistency to his thoughts and actions. Springer retained an interest in the university and a deep appreciation for the experiences and relationships that he had enjoyed there. From a perspective colored by time and distance, he would one day recall his years at the university as "the happiest part of my life—those joyous and careless days that will come back to us—never again."[13]

The Young Attorney

Having considered and rejected science as his vocation, Springer turned to the law as his chosen profession. He faced the practical necessity of making a living, and it was his desire to live in a comfortable style and to be regarded by others as a successful man. In his view, success was measured at least partly in financial terms.

Springer was also looking to the future, and a tentative life plan was emerging. An attorney's earnings would provide a stake for the kinds of investments in which real money could be made. If he could somehow attain financial independence, Springer might yet realize the ideal of a life devoted to science. As a first step, he turned his attention to the study of law.

Springer's legal training reflected both the traditional method, "reading law" in the office of a practicing attorney, and the emergent—formal education in a university law department. He was accepted to read law in the office of Henry Strong at Burlington. Strong was general counsel for the Chicago, Burlington and Quincy Railroad, and he was highly regarded in legal and business circles. He was later engaged in legal practice and business in Boston and Chicago, serving briefly as president of the Santa Fe Railroad.

It was Springer's good fortune to be associated with Henry Strong during the several years of his residence in Burlington. As Springer's mentor, Strong dispensed not only his considerable knowledge of the law but also his wisdom concerning the profession. He also critiqued the leading attorneys of the day, pointing out truly useful characteristics as well as qualities that were of little real value.

In 1868 the Iowa legislature appropriated funds to relocate the law school originally established in Des Moines and incorporate it as a department of the state university.[14] Springer returned to Iowa City for the spring term in 1869. According to his own recollection, Springer "took the senior term at the Law School of the University and the examinations for the full course, but not the degree because not long enough in residence."[15] At this time, the full course in law was taught in one academic year, and Springer attended for the second term only.

Upon his return to Burlington, Springer was given work in the office of his mentor. The intention was for him to assist Strong in the preparation of cases while studying for the bar examination. However, Springer soon was compelled to take a more active role, owing to the unexpected absence of Strong. In order that he might argue a case coming before the circuit court, he was given a special examination for admission to practice in Iowa. According to notes recorded in his pocket diary, Springer was "admitted to the Circuit Court by Judge Power" on August 5, 1869, made a successful debut in legal argument on August 6, and spent most of the next week in court finishing the case in Strong's absence.[16]

According to Springer's recollection, he was opposed by two leading attorneys of Burlington, but won his first case. As a result of his performance, Strong took him as a partner, putting him in charge of a number of railroad cases.[17]

In 1870 Springer was appointed assistant to the district attorney for the First Judicial District of Iowa and was placed in charge of criminal matters in Des Moines and Louisa counties. Some of the cases he handled were heard on appeal by the supreme court of Iowa, and at least two resulted in the establishment of new precedents in Iowa law.[18]

One notable case was first tried in the district court of Des Moines County in April 1871. The case, involving two men and a grudge, resulted in the indictment of Aquilla Lawrence for second-degree murder. In the district court, Springer won a conviction for manslaughter and the praise of the *Burlington Daily Gazette,* which furnished readers with a critique of the young lawyer's performance.

> The argument in the Lawrence case was commenced this morning. Mr. Frank Springer, assistant counsel for the state, made the opening address for the state, occupying about two hours in its delivery.

Frank Springer, c. 1870. Photo by Montfort and Hill Studio. Courtesy of Museum of New Mexico, Neg. 45480.

> Mr. Springer is the son of Judge Springer, for many years our district judge. He is a young man, but has displayed on many occasions an ability and maturity of judgment that would do credit to a much older head. His general knowledge of the law is good, and he has evidently given much thought to the subject. He handles facts with considerable skill, but has the fault of repetition, which he will correct as he grows in experience. He made a good argument this morning, and it is a flattering testimony to his ability that he was permitted to open the case, when he had so many able lawyers opposed to him.[19]

Springer won a conviction in the district court and his case withstood an appeal to the Iowa Supreme Court.[20]

While living in Burlington, Springer also maintained an active interest in scientific matters, exploring fossil-rich formations in the limestone bluffs along the Mississippi and remaining in touch with Orestes St. John, whose work for the Geological Survey occasionally brought him to Burlington. Even as he prepared for a career in law, Springer found time for sojourns to look at fish teeth or hunt for fossils. He also began a fruitful association with his eventual scientific collaborator, Charles Wachsmuth of Burlington, enlarging his own understanding of the Crinoidea and enjoying the use of Wachsmuth's fossil collection and library.

In 1872 Springer was able to spend several weeks with Louis Agassiz at Harvard. Having settled on a legal career, Springer had reluctantly concluded that he could not do justice to his profession and to the scientific studies that he so enjoyed. He expressed himself to Agassiz and was grateful ever after for the older man's advice. "Do not abandon the study," Agassiz told him. "Keep it up if you do but little. You will find it a resource in later years the value of which you cannot estimate now."[21] Agassiz's advice was sound and impressive, prompting Springer to retain his fossil collection and continue accumulating materials that would one day constitute a substantial scientific library.

A tragedy befell Springer's family in Columbus City when, in March 1872, his younger brother Warren drowned in the Iowa River while hunting near his home. Upon learning of the accident, Frank Springer and his father hurried home from Burlington to join in searching for the body, which was found downriver four days later.

While living at Burlington, Springer shared bachelor quarters with his father. Following his appointment as collector of internal revenue, Francis

Springer had remained in residence at Burlington while his wife kept the family home at Columbus City. At one point Frank Springer wrote to enlist his mother's support in persuading his father to buy a new house, explaining that the present quarters leaked and he was tired of being cramped.

The son, however, would not need a residence in Burlington much longer. By the end of 1872, he was giving serious consideration to an offer of employment that would alter the course of his life and career. Springer stood to do well in Burlington, but he was also aware of extraordinary opportunities in the developing western territories. When an opportunity came his way, he gave it more than passing attention. Mindful, perhaps, of the adage "nothing ventured, nothing gained," he ventured.

CHAPTER 2

New Mexico

By the end of 1872 Frank Springer had made a good beginning in his career as an attorney. He had attached himself to one of the outstanding members of the Iowa Bar in his mentor and partner, Henry Strong, and he had begun to make a name in his profession. Although science had been relegated to the status of an avocation, Springer was in an area that was rich in fossil formations and where he could enjoy the fellowship of men who shared his interest.

Springer appreciated his advantages and was intensely loyal to his family, his university, and his friends. At the same time, he nurtured ambitions beyond a career in Burlington. He wanted to attain a level of success that would one day allow him to set aside the daily responsibilities of his law practice and pursue scientific and other interests. His opportunity came in the form of a proposal from his good friend William R. Morley, who had gone west after leaving the state university, and who now was managing affairs of the Maxwell Land Grant and Railway Company in New Mexico.

During the recent war, Morley had helped restore the war-ravaged railroad line from Chattanooga to Atlanta; in so doing, he found a career. Upon his return to Iowa, Morley enrolled at the university to study civil engineering.[1] In the spring of 1869 he headed west, gaining employment with the Kansas Pacific, then laying rails in Colorado and Wyoming. His competence and courage attracted the attention of William J. Palmer, managing director of the Kansas Pacific Railroad.

The Maxwell Land Grant

While Palmer and Morley drove the Kansas Pacific toward Denver, several hundred miles south, in the territory of New Mexico, Lucien Bonaparte

Maxwell was contemplating a sale of his extensive land holdings. Maxwell, a former scout and buffalo hunter, an associate of John C. Frémont and Christopher "Kit" Carson, had come into the old Beaubien and Miranda Land Grant by marriage and through a series of purchases that allowed him to consolidate ownership of the grant over a period of years beginning in 1858.

The grant was originally made to Don Carlos Beaubien and Guadalupe Miranda, residents of San Fernandez de Taos, on January 11, 1841, under authority of Manuel Armijo, governor of Mexico's northern province. Personal and political motives doubtless entered into the allocation of land grants, but their essential purpose was to provide for defense against hostile Indians and foreign powers on the northern frontier of New Spain and then of Mexico. Grants also stimulated development of lands previously unsettled by people of Spanish or Mexican origin.

The governor's action was protested almost immediately by Padre Antonio José Martínez of Taos, but survived the challenge and was later confirmed by the departmental assembly. This was but the beginning of a long history of bitter conflict concerning the legal sufficiency of actions taken by the Mexican and U.S. governments, as well as the size and legal boundaries of the grant.[2] Matters were complicated by the transfer of sovereignty over the area from the government of Mexico to that of the United States, and by the slow, indecisive, and sometimes self-contradictory manner of the U.S. government in moving to quiet Spanish and Mexican land titles.

An accelerating influx of settlers, increasing tension over land titles, and the frenzy precipitated by the discovery of gold on Baldy Mountain in 1866 complicated efforts to manage the estate. Scarcely ten years after acquiring the grant, Maxwell was considering options for disposing of the property. In May 1869, a group of investors headed by Colorado senator Jerome B. Chaffee acquired an option to purchase the grant. Negotiations ensued, Chaffee enlisting the participation of a group of British capitalists. By deed of sale dated April 30, 1870, the grant was conveyed to its new owner, the Maxwell Land Grant and Railway Company. Needed capital came from the sale of securities to British and Dutch investors.

Among prominent American citizens forming the board of directors for the new company was William J. Palmer, who was also elected president of the board. Palmer employed Morley as engineer and surveyor for the company. Palmer soon yielded the presidency to John Collinson, leader of the English syndicate and majority stockholder in the enterprise, and Collinson installed another Englishman, John Langham Reed, as vice president and chief executive officer. Reed, in turn, succeeded to the presidency of the struggling company; when he withdrew, the trustees elected the company attorney, a rising political and commercial force in the territory, Stephen B. Elkins.

Morley took up residence in Cimarron and served in his original capacity until November 11, 1872, when, at the urging of one of the directors, Henry M. Porter, he was appointed the company's executive officer.[3] By contract between Morley and Elkins, Morley was to receive an annual salary of $4,500, the services of a clerk, the use of three company horses and a buggy, and two rooms and a kitchen in the "Cimarron House," as the old Maxwell mansion was called.[4] Family tradition holds that Morley accepted the position, in part, out of a desire to provide a suitable home for his beloved fiancée, Ada McPherson. Morley and Ada were married in January, 1873, and took up residence in the old Maxwell home. Morley was shortly awarded the title of vice president of the Maxwell Land Grant and Railway Company.

As executive officer, Morley inherited a morass of financial and legal problems, owing chiefly to the greed and ineptitude of the previous management. The company labored under a staggering debt, and its title to the property was widely challenged by settlers who refused to pay rent or purchase the lands they occupied. Similar problems were evident in the Baldy Mountain region, where gold seekers refused to pay for the privilege of mining on property claimed by the Maxwell Company. The presence of Ute and Apache Indians, while providing some actual and potential sources of income, made it difficult for the company to attract investors.[5] Morley inherited these difficulties, along with a newspaper—a necessary adjunct to the company's promotional objectives, but one to which he could hardly give adequate attention.

Melvin W. Mills and Thomas B. Catron were handling a large volume of legal business for the company, but Morley was eager to have his school friend Frank Springer assume the role of local attorney. It has been suggested that Morley's enthusiasm for Springer had to do with his increasing distrust of Elkins and Catron. They were the presumed leaders of the emerging "Santa Fe Ring"—a collective of attorneys, politicians, and moneyed interests that appeared to exert extraordinary control over political and legal affairs of the territory, to the personal advantage of the principals.[6]

One of Morley's first acts as manager of the property was to urge Springer to join him in New Mexico Territory. Springer would assume a major share of the company's legal business and help manage and edit the *Cimarron News*.

To the Southwest Frontier

Springer considered Morley's proposal and found it sufficiently intriguing to justify abandoning his legal practice in Burlington. His departure was delayed several days due to transportation difficulties out West. Horses in the Rocky Mountain region had been suffering with the "epizootic." As a result, regular service on the stagecoach route had been suspended.

Springer was able to start from Burlington on January 26. His arrival in Cimarron was much anticipated, as Morley on January 27 wrote company president Elkins, "Springer will be here as soon as coaches run which for the sake of our business I hope will be soon."[7] The trip took eleven days and involved travel by rail to Omaha, Cheyenne, and Denver; narrow gauge train to Pueblo; and stagecoach to Trinidad, Colorado. The final leg of the journey, over Raton Pass and across the plain to Cimarron, was made in a spring wagon drawn by a team of mules.

At Cheyenne, Springer took advantage of a delay to advise his mother of his progress. He complained of the monotony of the Nebraska prairies but looked forward to a more interesting ride through Colorado, where mountains would be in view all along the way. He also reported seeing two large herds of antelope and a camp of Indians—twenty or so lodges on a hillside.[8]

Of the growing city of Colorado Springs, Springer was moved to remark, "It is the prettiest little town I have seen, and as it is at the entrance to the most celebrated localities for tourists in Colorado, it is bound to become a place of great attraction if not importance." Springer was impressed with the grandeur of the Rocky Mountains, but had some difficulty in adjusting to the immense scale of the landscape, telling his father: "I spoke of Colorado Springs as being at the base of the mountain. So it seemed, and it occurred to me that a journey to the mountain would be but a few minutes walk, and up it only a few hours. When I learned that it was 12½ miles, in a direct line, from the village to the mountain proper, and a two days labor to make the ascent and return—I congratulated myself that I had not been fool enough to express myself on the subject."[9] At Pueblo, Springer was delayed for several days. Although regular stage runs had resumed, the temporary suspension of service had left the company with a backlog of mail to be delivered down the line, so passengers were not being taken on the southbound route.

On February 3, Springer obtained passage from Pueblo, but the stage line advanced his cause by some eighty miles, only to leave him stranded at Trinidad. Forced to lay over a day before catching a ride in a wagon bound for Cimarron, he passed the time by climbing a peak near the town, an experience he found to be "exhilarating and intoxicating in the highest degree."[10] Springer enjoyed "a most delightful ride" from Trinidad to Cimarron—a distance of some sixty-five miles, taking a day and a half in a mule-drawn spring wagon. He arrived in Cimarron on February 6. The next day he wrote to inform the home folks of his arrival.

"In Cimarron I have been most agreeably disappointed," Springer wrote. "I had become somewhat disgusted with the brown mud houses, and expected of course to see them here also. But to my surprise I found a most pretty little village, beautifully situated on and near two nice little streams, with houses clean and whitewashed, and some of them two story buildings of stone."[11] Springer was given temporary lodging in the old Maxwell mansion, but was shortly to have his own quarters and a law office near the newspaper office. Within days of his arrival in Cimarron, he had purchased a parcel of land that would form the nucleus of a cattle ranch.

A Start in the Territory

Springer had expected time in which to familiarize himself with the territorial laws of New Mexico before commencing active legal practice, but the condition of the Maxwell Company was such that his services were needed immediately. Morley wrote to the company's Santa Fe attorney, Thomas B. Catron, asking that he bring Springer up to date on pending legal matters.[12] Among other duties, Springer was to make out suits of ejectment against a number of settlers considered by the Maxwell Company to be occupying lands illegally.[13] Having sized up the situation, Springer told his father:

> I find I am likely to be plunged head over ears in law business at once. The company is badly mixed up in litigation growing out of the gross mismanagement of the officers before Morley came in and although it

> was stated at first that I would not be expected to attempt much litigation the first year, affairs are in such shape that I shall have to do it, and I am going in to win if at all possible. I have on hand half a dozen suits involving from two to ten thousand dollars, which I am preparing for court and will either try or assist in trying. My library is not here yet, though it is on the road, and I think it will be here in a day or two. It is rather uphill business practicing law with one book, and that the most miserable compilation of absurd and crude statutes you ever saw.[14]

Springer also reported that newcomers could settle in the region without fear of molestation from the Indians—Utes and Jicarilla Apaches who still inhabited the area—but that the company was going to make an effort to have them removed from Colfax County.

The affairs of the Maxwell Land Grant and Railway Company and his own pursuit of prosperity were enough to keep Springer occupied during his first months in Cimarron. He took on his share of editorial duties for the *Cimarron News,* tended to such legal matters as were referred to him by Morley, and placed a business card in the newspaper advertising his services as attorney and counselor at law, with offices on the north side of the plaza.

Springer's removal to New Mexico brought about a temporary suspension of his scientific studies, but he sent instructions home concerning the care of his fossil collection, and he wrote Charles Wachsmuth, telling of his life in the Southwest and his fond recollection of the time in Burlington. "I need not tell you how much I enjoyed the time we spent together," he wrote. "You could see that for yourself, and I look back on the trips we used to make among the rocks as among the most pleasant hours of my life. If I was not entirely engrossed by business, the remembrance of them would make me a little homesick. But I came out here to make money, and I think I shall succeed, and also do it in much less time than I could have done in Burlington."[15]

Springer gave an account of the country, describing a visit to the Taos Valley and his impressions of the Pueblo Indians, the Penitentes, adobe buildings, the Rio Grande Gorge, and the Ojo Caliente hot springs. His response was, in many ways, what might have been expected of a newcomer whose cultural frame of reference was eastern Iowa. The Penitente sect struck him as "a relic of barbarism." Springer had heard tales of self-torture with thorns and soapweed whips, and had seen the visible evidence of massive wooden crosses on lonely hillsides. "The Penitentes are a secret organization, which in some respects resembles the Jesuits," he wrote. "They stand by each other in everything, in crimes as well as anything else."

Springer later told Wachsmuth of his experience panning for gold among the placer mines of the Baldy country, where he was able to wash five dollars in gold in an about an hour. Of his new routine, he remarked, "I do most of my traveling horseback now, and have become accustomed to it, so that a ride of 40 or 50 miles a day does not tire me much."[16] Springer referred to worsening economic conditions and admitted that times were hard, "though not more so proportionally than in other parts of the country." A pilgrim in the territory, living a spare existence with few amenities, he confided, "When I have made my fortune, perhaps I may come back to Iowa to live. I would much prefer it, were money as easily made as here."

The Maxwell Company in Distress

Enormous difficulties beset the Maxwell Land Grant and Railway Company as a result of careless management and profligate spending on the part of the previous administration. To his father, Springer described the situation in plainest terms, placing blame squarely on the company's former managers. "Over a half million dollars have been squandered here, prior to the incoming of the present management, with practically nothing to show for it," he complained. "And we are even now embarrassed by a load of doubtful lawsuits and contracts which none but idiots would ever have entered into."[17]

The company's problems were compounded by a panic that started in the country's financial centers, but soon curtailed investment in all sections. Signs of an impending collapse were discernible throughout the summer of 1873, as interest rates and inflation rose. The crisis came in September, when several large banking establishments suspended business. Stocks and other securities plummeted, and with them the confidence of badly shaken and, in some instances, badly damaged investors.

Among other consequences, the panic halted a boom in construction of western railroads. Out West, those most directly affected by events in the financial centers were justifiably alarmed, but others were able to bear the news philosophically. Springer recognized the difficulties for his region, which pinned great hopes on the prospects of a railroad, but he remained optimistic, telling his father, "Times are very hard here, money scarce and collections difficult to make. I have great hopes of the success of our railroad in the face of the panic. But I think it will be the only railroad built in the West for some time."[18]

The Maxwell Company's letterhead already indicated not one, but three railroads connecting Cimarron to the rest of the country. These were but the wishful thoughts of the company's promoters, graphically expressed on a map indicating the property's still much-disputed boundaries.

In spring of 1874 Morley made a game attempt to gain financing for a railroad, traveling to the Netherlands to sell the proposal to the company's stockholders. Morley had gone to Washington to confer with Stephen B. Elkins, who did double duty as New Mexico's territorial delegate to Congress. While Morley was in the East, a cable arrived from Amsterdam requesting that Elkins come at once; as Elkins was otherwise occupied, it fell to Morley to make the long transatlantic voyage.

On the eve of Morley's departure, Elkins wrote to his employee in New York, wishing him well on the journey and promising to secure the company's title to the property. "I can and will settle the question of title," he assured Morley. "I will have Drummond's order revoked beyond doubt and a bill passed by which we can adjust our title without delay."[19] Willis Drummond, commissioner of the General Land Office, had recently issued an order declaring the Maxwell Company in default on obligations relative to its efforts to secure legal title to the property and providing for extension of the public surveys over the grant. This was widely understood to mean that the entire Maxwell claim, with the exception of approximately ninety-seven thousand acres, was to be declared public domain and opened to settlement under auspices of the General Land Office.

Elkins instructed Morley to assure the company's creditors that "everything is safe," but he also took the occasion to press his claim for some eleven thousand dollars owed him for services. "I have written [the Dutch investors] this must be paid or I will resign and I will," he fumed. "I cannot afford to pay Catron and Waldo as I am unless I am paid—besides just at this time I am doing important service for the Company and you know it."[20] Scrawled on letterhead of the U.S. House of Representatives, Elkins's missive was an uncommon example of official candor, since he was in fact doing important business for the Maxwell Company in Congress, attempting to annul the Drummond order and secure a favorable resolution of the title dispute.

Morley did not obtain the needed funding and the company's long quest for railroad access to the grant was again frustrated.

Frontier Editor

Among Springer's duties in the service of the Maxwell Land Grant and Railway Company, he was expected to take on publication of the *Cimarron News,* a weekly appearing each Saturday. The paper carried the usual fare of local news, personal notes, national and world events, and advertisements. As with frontier papers generally, the *Cimarron News* served as an organ for boosting the region's prospects and encouraging immigrants to come, invest in property, and settle. Expressions concerning the prospects for railroad construction and other favorable developments were unfailingly optimistic. The *News* served the further purpose of advocating the position of the Maxwell Land Grant and Railway Company in its dispute over ownership of the grant property, while making some effort to demonstrate fairness in its reporting of the conflict with settlers.

Of further concern was the continuing presence of Ute and Jicarilla Apache Indians. They were not considered dangerous, but were regarded as an element likely to retard development. Springer took pains to assure his mother that the local Indians were peaceable, but he was in sympathy with efforts to have them removed to reservations. The *News* frequently commented on the Indian problem. In January 1874, the paper carried an editorial on drunkenness among the Indians, reporting on recent incidents on the property of M. M. Chase, a rancher of the Ponil region, calling the whole situation "an unbearable nuisance" and elaborating as follows: "When the Indians are sober they are very peaceable and easy to get along with, but when drunk they are still less reasonable than drunken white men. All the brutality of their nature is aroused, and they will do anything. This condition of things should not be permitted to last long, for men who live out of town dare not leave home without being haunted by a horrible fear that their wives and children may be set upon and abused, beaten or perhaps killed, by a party of drunken Indians."[21]

Springer's attitude concerning the Indian presence doubtless reflected the Maxwell Company's position and that of many other settlers from the States, but the sentiment was not universal. Rev. Franklin J. Tolby, a Methodist minister working in the area, believed that the Indians should be permitted to remain in the region. To this end, he attempted without success to have a reservation established on the grant. Jicarilla Apache leaders had repeatedly expressed the desire of the people to remain in their familiar homeland near Cimarron, Abiquiu, and Tierra Amarilla, but the U.S.

government proved no more helpful to the Indians than private claimants had been. Various attempts were made to resettle the Jicarillas, but government indecision and factionalism within the tribe kept the matter unsettled until 1887, when Jicarilla leaders agreed to locate on a reservation west of the Chama River.[22]

By June 1874, Springer was able to turn the management of the *News* over to J. H. Koogler and William H. H. Werner. Springer signed off in a brief card to readers, commending the new management and assuring readers that they would find the change beneficial.[23] Springer was not entirely finished with responsibilities relating to the *News* as he may have supposed, but he could—for a while, at least—turn his attention to other matters. Of immediate interest were the many opportunities to invest in the region. Land and cattle could be had at favorable prices, and opportunities for new industry and business development were many. Most of Springer's investments were in land, but he also purchased a part interest in a sawmill.

Springer could not devote his full attention to business matters, for in the summer of 1874, he had a new worry. His mother's health had been deteriorating steadily. Although the railroad had not yet reached Cimarron to provide convenient access, the family concluded that her best chance for recovery lay in the arid climate of New Mexico. In June, Nancy Springer and her daughter, Nellie, traveled to Cimarron. His mother was weak, but through the summer, Springer saw hopeful signs that her health was improving. "About the middle of November, my mother had a severe attack of Pneumonia, which in her weak state she was unable to resist, and which terminated fatally in a few days," he recalled. Springer accompanied the remains home and stayed several weeks, visiting family and seeing associates at Burlington.[24]

From time to time Springer thought of returning to Iowa to live. His mother's death and the extended visit home surfaced the issue again. Springer gave the matter serious consideration, but as he weighed the likely consequences, his confidence in the decision to go west was affirmed. He told Wachsmuth, "I am offered some inducements to return to Burlington for good, but scarcely think I shall do it at present. I think I can do better west and I intend to devote the next few years of my life to accumulating a sufficient fortune to enable me to gratify my desires and follow my favorite pursuits without embarrassment as to finances. I have confidence in the west and believe that it is the most favorable place for me."[25]

The arrangement by which Springer and Morley were liberated from day-to-day responsibility for the newspaper did not last long, for by November 1874, the *Santa Fe New Mexican* was reporting that the Cimarron paper had ceased publication and would not reappear under its previous management.[26] Springer and Morley again found themselves in need of an associate to whom they could entrust the paper's day-to-day affairs. They located such a man, they believed, in Will D. Dawson. The new arrangement constituted, in effect, a merger of the defunct *Elizabeth City Railway, Press and Telegraph* and the *Cimarron News*. On behalf of Springer and himself, William R. Morley contracted with Dawson, editor of the late Elizabethtown paper. Morley enumerated Dawson's duties, specified the paper's editorial policy, and stipulated that it was to be styled the *Cimarron News and Press*. Editorial oversight was reserved to Springer and Morley, with one of them, unspecified for the moment, to assume this responsibility. On land grant questions, the paper was to maintain a neutral position, "publishing facts on either side but not advocating either side in Editorials."[27] Morley, Springer, and Dawson all appeared jointly in the new paper's heading as "editors."

In less hopeful moments Springer and Morley might have predicted that the arrangement would not go smoothly. Dawson had already shown himself an opponent of the Maxwell Land Grant and Railway Company, publishing an unfavorable article from the *Denver Mirror* in the *Elizabeth City Railway, Press and Telegraph*, alleging fraud in the matter of the grant's boundaries.[28] Dawson was also a friend of Melvin W. Mills, a Colfax County lawyer with strong ties to the Santa Fe Ring.

A Question of Title

From the time Lucien Maxwell sold the grant and investors with purely pecuniary interests took over, tension had been building. At issue were the status of the grant, the company's efforts to enforce its claim to the property, and the uncertain rights of settlers—farmers, ranchers, and miners who worked the land but who could not produce satisfactory title to their claims.

The company's claim to land totaling nearly 2 million acres rested on a Mexican grant, the boundaries of which were described and mapped in a manner leaving some precision to be desired. By an act of Congress adopted

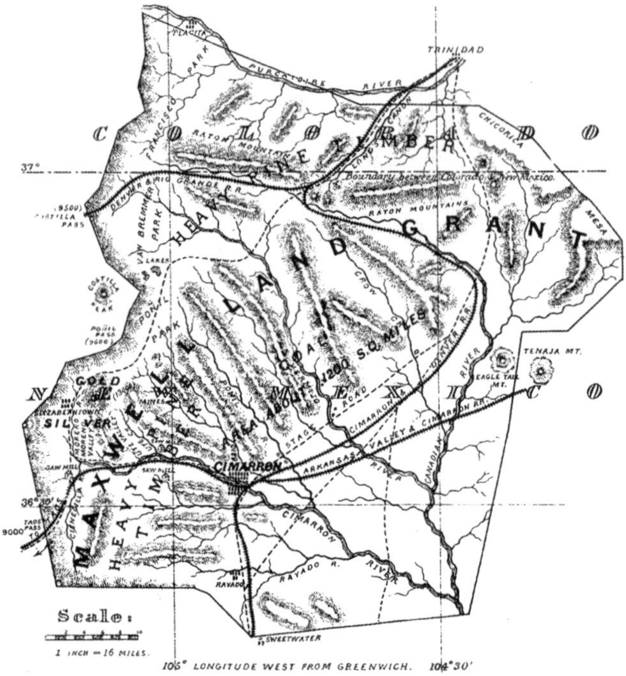

Maxwell Land Grant according to the Griffin Survey of 1869, from Maxwell Land Grant and Railway Company letterhead, c. 1873. A later survey excluded the lobe extending from the north end of the eastern boundary, but the grant was otherwise confirmed substantially as shown. Courtesy of CS Cattle Company.

June 21, 1860, the U.S. government appeared to confirm the grant, regardless of size, but its size did become an issue. In 1869, as Maxwell was contemplating sale of the grant, an official survey based on the boundaries described in the original Mexican documents was authorized. Deputy U.S. Surveyor William W. Griffin commenced the survey, but on December 31, 1869, it was suspended on order of Jacob D. Cox, secretary of the interior, who ruled that the original grant to Carlos Beaubien and Guadalupe Miranda—the grant now owned by Maxwell—could not exceed 97,000 acres. Cox's ruling was based on a Mexican colonization decree limiting grants to individuals to eleven square leagues; hence two applicants could claim no more than twice that amount.

This ruling occurred at a delicate time, as a sale of the property was pending. Confident that the grant, as described in the Mexican documents, had already been confirmed by Congress, and confident that the claim could be sustained in litigation, the buyers proceeded with the transaction. Griffin was contracted to complete the survey on behalf of the purchasers. The survey was filed with the General Land Office and the sale went forward. Chaffee and his two partners reached agreement with the Maxwells, then assigned their option to purchase to the Maxwell Land Grant and Railway Company—an English syndicate. The sale went through and the new owners took possession of the property.

The decision of Secretary Cox in attempting to limit the company's claim to 97,000 acres, combined with the sale to foreign interests, contributed to a reluctance on the part of many settlers to comply with demands of the Maxwell Company that they pay up or vacate their claims. Commissioner Drummond's 1874 order further compounded the company's woes.

Despite these difficulties, the Maxwell Land Grant and Railway Company hewed to the belief that its title would be upheld and pressed its claim accordingly. Morley sought to enforce property rights claimed by the company, but in the absence of an unequivocal ruling on the part of the government, recalcitrant settlers refused the company's demands. By June 1874, Morley had for some months been running a notice in the Cimarron paper, advising settlers that the company would not suffer unlawful occupation of its lands. Unauthorized parties were cautioned that "unless they quit and surrender the possession of said lands and premises, they will be treated as trespassers and proceeded against according to the law."[29]

Resistance to this point had been largely rhetorical, with confrontations taking place in spirited newspaper exchanges, in the territorial courts, and in heated conversations between settlers and functionaries of the company. A "Squatters Club" had been organized to advocate for the interests of the settlers. In the early months of 1875, tensions increased. The conflict between Maxwell authorities and the "squatters" was responsible for much of the unease, but men on both sides of that issue found themselves united in opposition to the heavy-handed manipulation of the Santa Fe Ring. A growing wariness among the men of Colfax over the increasing power of the Ring led to a rift between Morley and Springer on the one hand, and Catron and Elkins on the other.[30] Elkins's administration of company affairs had ended with his resignation as president in the fall of 1874; from there, the relationship had deteriorated.

Springer became directly involved in the land grant controversy, first as the likely writer of a lengthy treatise on the legal status of the Maxwell Land Grant and later as a candidate for the territorial legislature, representing

Colfax County. In May 1875, the *Cimarron News and Press* published a long, elegantly written, and highly detailed statement that, while assuming an objective posture, advised that the company's claim to the grant was secure and supported entirely by legal precedent.[31] The piece was signed "L.E.X.," a signature likely identifying Frank Springer as the article's author.[32]

As a candidate for the territorial legislature during the summer of 1875, Springer was called upon to address several matters of interest to citizens of Colfax County, including his position on "the persons known as squatters, who have located upon the lands claimed by what is known as the Maxwell Land Grant Company." Springer minimized the differences of the company and the settlers. Of paramount importance, he emphasized, was the settlement of land titles, including the Maxwell claim. With this accomplished, he concluded, settlers might purchase their lands on comparable terms from the government or from the company, and get on with the business of developing the land and securing a prosperous future.[33]

In the view of Victor Westphall, Catron's biographer, Springer managed to finesse a delicate situation during the early years of his association with the Maxwell Land Grant and Railway Company, remaining loyal to the company while appearing at the same time to advocate for the interests of the settlers.[34] Springer's growing role in affairs of the Maxwell Company would later make it difficult for him to maintain such a precarious balance.

Springer lost the election for legislative representative from Colfax County to Melvin Mills. The general election was held on September 6, 1875, and on September 13, the *Santa Fe Daily New Mexican* reported Mills the conclusive victor by a margin of 214 votes over Springer.[35] Springer may have emerged from this experience more fully persuaded of the Ring's inordinate political power, and with a giant burr under his saddle.

Storm Clouds

New Mexico had recently acquired a new territorial governor in Samuel Beach Axtell, the late and largely unlamented governor of the territory of Utah. The *Daily New Mexican* welcomed the new appointee, characterized by the paper as a person "manifestly intelligent, pleasant and entertaining in conversation."[36] Frank Springer's irritation concerning the practices of the Santa Fe political machine would soon come to focus on the new governor, a man Springer found to be mildly entertaining, but neither intelligent nor pleasant.

Meanwhile, Springer's associates in resistance to the Santa Fe Ring, Morley and Rev. F. J. Tolby, had been exercising their rhetorical gifts as correspondents for the *New York Sun*. The two men were widely assumed to be the authors of a series of unsigned articles that purported to expose the corruption of the Ring. Tolby was no friend of the Maxwell Company, but he and Morley found common ground in their mutual contempt for the Ring. Tolby made no secret of his opposition to the Ring, publicly voicing his criticism to residents of Colfax. An article in the July 5, 1875, edition of the *Sun,* headlined "The Territory of Elkins: The Petty Despotism That Is New Mexico," informed readers that the population of an entire territory was at the mercy of "a firm of sharp lawyers." Elkins and Catron were identified as leaders of the Ring, and they, along with Judge Joseph Palen and others, were accused of manipulating elections and court proceedings to benefit themselves.[37] Understandably, hackles were raised on the part of those who were implicated in the *Sun* articles.

Tensions were further heightened by occasional outbreaks of violence. In June 1875, Francisco "Pancho" Griego, a leader among the Hispanic people of Colfax County, shot and killed two soldiers from Fort Union in a saloon brawl at the Saint James Hotel in Cimarron. It was widely believed that powerful elements of the Santa Fe Ring intervened to spare Griego the inconvenience and possible embarrassment of facing criminal charges in exchange for his political support. Frank Springer was among those who subscribed to this notion, as indicated in a deposition he gave on the troubles in Colfax County a few years later.[38]

Another event that fueled the controversy involved Morley's mother-in-law, Mary Tibbles McPherson. A woman of strong convictions, Mrs. McPherson was in New Mexico in the spring of 1875 to visit her daughter and see a new grandchild, but upon hearing stories of corruption in the territory, she became indignant and undertook to address the matter. She fired off a letter to Washington, complaining of the state of affairs in New Mexico and blasting territorial officials supposed to be members of the Santa Fe Ring. Thinking to head off trouble, Mrs. McPherson's daughter hurried to the post office to retrieve the letter. She was observed doing so, and was reported to the authorities for theft of the U.S. mail.[39]

When it became apparent that Mrs. Morley was to be formally accused in the matter, Frank Springer made a trip to Santa Fe, seeking to forestall an indictment. Springer did not succeed, but the contact resulted in a hardening of the line between Ring sympathizers and the contrarians of Colfax. Springer later came into possession of a letter from Melvin Mills to Robert Longwill, telling of Springer's mission in Ada Morley's behalf. Reported Mills, "We are having great times here—the delegation from Cimarron is here and I think we have the best of everything. Springer gets a damn cold reception from everybody. Mrs. M. will be indicted without fail."[40]

Among the responsible authorities was Thomas B. Catron, once and future figure in affairs of the Maxwell grant, but now U.S. attorney for the territory and an enemy of William R. Morley. At the urging of the Cimarron postmaster, Mrs. Morley was indicted and a warrant issued for her arrest. The indictment was widely interpreted as a vindictive act on Catron's part. In apparent consideration for her pregnant condition, the warrant for Mrs. Morley's arrest was not immediately served. The matter was eventually allowed to drop, but the incident inflicted considerable damage on Catron's reputation and, in Westphall's view, hastened the end of his tenure as U.S. attorney for the territory of New Mexico.

Springer's first two and a half years in the territory had been eventful, to say the least, but his experience to this point was as nothing compared to the turmoil in which he and virtually all residents of Colfax County were soon to be embroiled. Had Springer's university education included a more intimate acquaintance with the Spanish language, he might have expected that a frontier town named "Cimarron" would have a penchant for wildness. It had become a noticeably rowdy place already, but it was about to get a good deal wilder.

CHAPTER 3

The Colfax County War

Like a simmering volcanic mass, Colfax County was, by the summer of 1875, exhibiting signs of intense heat and pressure. Fault lines had appeared, dividing political and economic factions, and in increasingly spirited public dialogue one could discern the rumble preceding a major explosion. Contributory causes were numerous and varied—the presence of lawless elements so commonly found in frontier towns of the nineteenth-century West, a clash of races and cultures, an ongoing animosity between settlers and agents of the Maxwell Company, and increasing resentment of a political machine that seemed to manipulate affairs of the territory to its own advantage.

A new territorial governor, Samuel Beach Axtell, had been appointed in July, 1875. A native of Ohio and a former congressman from California, Axtell had already run into heavy weather during a brief stint as governor of the territory of Utah, stumbling into a dispute between the Mormons and a feisty anti-Mormon faction. President Grant could have turned the troubled executive out and left him to his own devices, but the New Mexico governor's office was vacant, so Axtell was given that appointment. Grant no doubt thought he was doing Axtell a favor, but the circumstances a new governor would face in New Mexico could hardly have been more difficult, and Axtell was not what anyone would call a master politician or administrator.

The Tolby Murder

All these factors increased the likelihood of violent conflict, but the proximate cause of the trouble that erupted along the Cimarron in the fall of

1875 was the murder of a reform-minded preacher who rose in opposition to the prevailing political and economic powers.

The preacher was Franklin J. Tolby, a young minister sent from the Northwest Indiana Conference of the Methodist Episcopal Church to serve the church's missions in New Mexico. Tolby's immediate connectional authority was the Reverend Thomas Harwood, longtime superintendent of the New Mexico missions. The circuit for which Tolby was responsible included congregations at Cimarron and Elizabethtown. Tolby had interested himself considerably in political matters and had become an advocate for the Ute and Apache Indians who remained in the area.

The known facts of Tolby's murder and the ensuing period of turmoil are recounted in at least a half dozen published accounts.[1] Springer's testimony, based on personal knowledge and contemporary hearsay, is contained in a lengthy deposition given in 1878 as a part of the report of Frank Warner Angel on the troubles in the territory of New Mexico, in Colfax and Lincoln counties.

Tolby had made no secret of his enmity toward the Santa Fe Ring. His role in producing articles for the *New York Sun* was widely suspected if not positively known. In any event, Tolby's public denouncements of the Ring made his position crystal clear to anyone who cared to listen. One of the objects of Tolby's wrath was Judge Joseph G. Palen, chief justice of New Mexico and district judge for the judicial district that included Cimarron. Palen had for some time been a controversial figure, owing largely to his supposed connection with Thomas B. Catron and Stephen B. Elkins in the Santa Fe Ring. It was believed in some circles that Judge Palen routinely decided cases in their favor in exchange for a tangible consideration.[2] Palen, in fairness, had as many supporters as detractors during his stormy judicial career, and in the view of Catron's biographer, Victor Westphall, the criticisms of the judge were largely unjustified.[3]

Springer's recollection of pertinent events immediately preceding Tolby's murder was as follows: "During the first week of September 1875, while the court was in session at Cimarron, Judge Palen and Tolby had a rather spirited altercation, the judge denouncing him for some remarks he had made about the court and its actions. Tolby immediately declared he would 'write up that judge so that 200,000 readers should see his record,' which led some to suppose that he might have been the author of some letters which had appeared in the eastern press reflecting on Palen, Elkins and others."[4] Matters were already tense in Cimarron, owing to the climax, on September 6, of a heated election campaign pitting forces of the Santa Fe Ring against residents who opposed the Ring, and to a recent and particularly contentious session of the district court.

On the morning of September 14, Tolby held services at Elizabethtown and then set out on horseback on his return to Cimarron. He rode through the wide expanse of the Moreno Valley, circled all around by heavily forested mountains, and over the ridge into the drainage of the Cimarron River. Between steep canyon walls, he followed the post road to a violent and untimely demise. On September 16 his dead body was found in the canyon, near the confluence of Clear Creek and the Cimarron River. Tolby had been shot twice in the back, but his money, watch, and other worldly goods had not been molested, and his horse and saddle were found in separate locations, each several hundred yards from the body.

Tolby, age thirty-three, was buried at Cimarron with full Masonic honors. Rev. Thomas Harwood, on an extended trip east, heard of the murder and hurried back to New Mexico, where he offered comfort to a grieving family. "Mrs. Tolby and her three children were literally crushed," he recalled.[5]

The news produced great excitement in Cimarron, especially among Tolby's friends and those who were in agreement with his views concerning the Santa Fe Ring. Speculation raged concerning possible motives for the murder. Some thought the killing was motivated by religious or racial prejudice, while others believed the deed was done to eliminate Tolby as a witness to an illegal act. Suspicion naturally fell on men associated with the Santa Fe Ring, who had reason to dislike Tolby and to want him eliminated for political as well as personal reasons.

Among those most interested in solving the murder was Tolby's assistant, Oscar P. McMains, a Methodist minister who had moved from Colorado only a few months earlier and who supplemented his modest stipend by working as a printer for the *Cimarron News and Press*. To many of Tolby's friends, the appointed agents of the law seemed slow to act—perhaps purposely so. Suspicion fell on Cruz Vega, who had been employed by the regular mail contractor—a supposed friend of the Ring—to carry the mail from Elizabethtown to Cimarron for one day, that being the day of Tolby's murder. Though not suspected as the trigger man, Pancho Griego was thought by some to have had a role in the murder.

Retribution and Repercussion

Impatient with the pace of the official inquiry, perhaps suspicious that Ring-controlled law enforcers did not plan to pursue much of an investigation, McMains took matters into his own hands. With the help of an intermediary, he lured Cruz Vega to a field along the Ponil River drainage near Cimarron on the night of October 30, 1875. There the suspect was confronted by a band of self-appointed investigators including McMains and Clay Allison. Some of the men with McMains reportedly had been drinking and may have been intoxicated. In the course of a rough interrogation, a lariat was placed around Vega's neck with the evident intent of extracting a confession or other useful information about the killing. This, at least, was McMains's intent. Desperate, Vega named Manuel Cárdenas as Tolby's murderer. By now the mob was beyond McMains's control. The free end of the rope around Vega's neck was tossed over a nearby telegraph pole and the helpless captive hoisted aloft briefly, then lowered to the ground. Having lost control of events, McMains apparently left the place. By morning Vega was dead.[6] A member of the coroner's jury later testified that Vega had been shot and bludgeoned, as well as hanged.[7]

The violent death of Cruz Vega shocked a region already reeling from the bitterness surrounding Tolby's murder and raised the appearance of an ugly racial confrontation, but the killing was far from over. On the night of November 1, Clay Allison shot Pancho Griego to death at the Saint James Hotel in a barroom altercation growing out of events surrounding the murder of Tolby and the lynching of Cruz Vega.

On November 6, Manuel Cárdenas, the man fingered by Vega, was arrested at Elizabethtown. Questioned by the authorities, he insisted that Vega had done the killing and named Florencio Donoghue, Melvin Mills, and Robert Longwill—all of whom were believed to be associated with the Ring—along with Pancho Griego, as persons who had hired Vega.[8] Cárdenas subsequently recanted his statements concerning the murder, but his retraction was disbelieved by many, including Frank Springer, who was convinced that the retraction was coerced by friends of the accused.[9] Cárdenas was formally examined on November 11, but as he was being taken from court after a lengthy hearing that lasted well into the night, he was shot dead by an unknown assailant. Springer reasoned that the friends of Tolby would have wanted Cárdenas alive, at least for the time being; his implied conclusion was that Cárdenas was murdered by forces of the Ring who wished to ensure his silence.[10]

From the grave, Cárdenas managed to make matters hot for the men named as conspirators in Tolby's death—Donoghue, Mills, and Longwill. Mills and Donoghue were taken into custody in Colfax County and questioned about the matter, but neither was ever brought to trial. Longwill fled to Santa Fe, reportedly eluding or outrunning a posse led by Clay Allison and Allison's brother, John, a deputy sheriff.

Meanwhile, Governor Axtell had requested that troops be sent to the area. A detachment of horse soldiers was dispatched from Fort Union, arriving in Cimarron on November 8. The *Weekly New Mexican* of November 16 led a story on the events in Colfax County with the headline "Anarchy at Cimarron" and reported that "[t]here seems to be an armed band of men in and about the town bent on violence, lawlessness and ruling to suit themselves." Said the *New Mexican,* "U.S. troops are on the ground and they will prevent violence and bloodshed if they have notice in time; but even their presence has not had the effect to disperse the mob who seem to have control of the prisoners and the administration of affairs."[11]

Opponents of the Santa Fe Ring viewed this action as unnecessary, provocative, and exemplary of the inordinate influence of the Ring. It was added to a growing list of grievances against the new governor. According to Springer's recollection, "The troops remained several days, and found nothing to do—found the people quiet and peaceable, and the law taking its course, but a very bitter and excited feeling among them on account of the murder of Tolby."[12]

McMains continued to stir the pot, organizing a citizens' committee to ensure that the investigation of the Tolby murder remained an issue. With McMains as chairman and Frank Springer as secretary, a meeting was held on November 10, 1875, for the purpose of verbalizing accusations against the Santa Fe Ring.[13] Springer and McMains would spend large portions of their lives locked in a struggle over the validity of the Maxwell Land Grant, but with the focus on Tolby and perceived excesses of the Ring, they were, for a time at least, allies in a righteous cause.

In the view of McMains's ecclesiastical superior, Rev. Thomas Harwood, McMains's clumsy efforts to see justice done in Tolby's murder were in no way helpful. Of the turbulent events following Tolby's death, Harwood wrote, "The excitement and danger were greatly intensified by the unwise efforts on the part of Rev. O. P. McMains who spent much time in trying to ferret out the case and bring the guilty parties to arrest." In Harwood's view, the chaotic situation induced by McMains's efforts may well have covered the tracks of persons otherwise culpable in the affair.[14] Judge Palen, who had played so prominent a role in the beginning of the Colfax County troubles, removed himself from further controversy by dying of natural causes on December 21, 1875.

Ring sympathizers were provoked when, on December 22, the *New York Sun* published yet another article on the trouble in Colfax County—this one furnished by an unnamed New Mexico correspondent, commending local efforts to solve the Tolby murder and restating grievances against the Ring. On December 31, Will Dawson, to whom Morley and Springer had entrusted management of the *News and Press,* annoyed the anti-Ring faction by criticizing the recent *Sun* article. Said the Dawson-edited *News and Press,* "We find another one of those lying, brutal letters in the New York *Weekly Sun* of the 22d inst., without signature, and purporting to come out of Santa Fe. The writer, whoever he may be, will be looked upon as a coward, a disturber of the peace, as well as a slanderer upon the good name and fair fame of this portion of New Mexico."[15]

Dawson declined to print the *Sun* article for his readers, but editorialized, "There can be none but the worst imaginable motive in any one who will persist in keeping alive the bad condition of feeling in this county which has grown out of distressing crimes, ending in blood and death." Some 153 local citizens signed a letter protesting Dawson's article and challenging the editor to publish their letter, along with the *Sun* article in question. This was eventually done, though not during Dawson's time as editor.

Unable to control Dawson and unwilling to assume responsibility for his editorial policy, Morley and Springer withdrew as joint editors with him on January 7, 1876.[16] Other citizens were more demonstrative in their expressions of displeasure. On January 19 a group of nocturnal visitors led by Clay Allison called at the *News and Press* office, where they proceeded to haul the press machinery out and dump it in the Cimarron River. This effectively ended Dawson's tenure as editor.

On January 28 Morley and Springer resumed the editorship of the *News and Press*. Having obtained another press, they put out a four-page paper in a reduced format. Under the headline "Richard Is Himself Again," the new management published a lengthy article stating the paper's editorial philosophy and advising readers concerning the new direction of the paper. "It is needless to say that this movement indicates a material change in the tone of the paper, as manifested in the last few issues, and a return to those principles, the advocacy of which during most of the past year has gained for it friends and patrons among the intelligent and thinking people throughout the whole territory; secured for it a circulation second to none in the territory; and made the *News and Press* the recognized organ of the people, the exponent of their rights, as against the abuses and outrageous laws and practices inflicted upon a long suffering territory by a servile and corrupt legislature."[17]

If Springer and others were annoyed at the notion of having U.S. troops sent to keep civil order in Colfax County, they were livid when in January 1876, the legislature saw fit to attach Colfax to Taos County for judicial purposes. This meant that sessions of the district court previously held in Cimarron would now be held in Taos. To the affront of being told, in effect, that the people of Colfax were not capable of managing their own affairs, was added the practical inconvenience of having to travel nearly sixty miles over a high mountain pass to attend sessions of the court. There was also some suspicion that the selection of juries in Taos was closely controlled by a local agent for the Santa Fe Ring. According to Springer, this bill, passed by the legislature on January 14, 1876, was taken up in the last day or two of the legislative session and without forewarning to the people of Colfax County.[18]

El Obispo

The new outrage was again laid at the feet of Governor Axtell, who averred that the action was necessary to maintain civil order in Colfax County. When the passage of the bill became known to the people of Colfax, Will Dawson—still editor of the *News and Press* at this time—telegraphed the governor, requesting that Axtell withhold his signature until a delegation from the county could meet with him. The governor's terse reply was, "Bill signed. S. B. Axtell."[19] On behalf of like-minded citizens of Cimarron, W. R. Morley, H. M. Porter, and W. O. Cunningham again tried to sway the governor in an exchange of telegrams on January 19 and 20. To the plea asking that the governor consider both sides of the matter, Axtell responded, "I see two sides. Broken laws, property wantonly destroyed and murdered men are on one side. I have no compromise to make with the other."[20]

The act of removing the district court from Colfax County moved Springer to action. He visited the governor in Santa Fe in February 1876, for the purpose of urging Axtell to reconsider the matter. Springer found the governor implacable:

> He spoke with extreme bitterness about the people of Colfax County, and informed me that he had visited every other county in the territory, and had intended to visit Colfax in its turn, but that now he should not do so. I asked and urged him to go to Cimarron and see the people and learn the facts for himself, but he positively refused, and said that he was fully informed about matters in that county and didn't need further information. He said there were bad men there and that he intended to have them punished or compelled to leave the country, if it took all the troops in New Mexico. He spoke particularly and repeatedly of one Allison and seemed very bitter towards him and mentioned him especially as one whom he intended to have indicted and punished, or compelled to leave the country.[21]

Upon Springer's return to Cimarron, he and several other citizens sent another message to the governor, entreating him to visit the community and meet face-to-face with the people. According to Springer, Axtell did not respond directly, but while in Las Vegas in early March, Springer encountered Benjamin Stevens of Albuquerque, district attorney for New Mexico's Second Judicial District, and was informed that Stevens would soon visit Cimarron for the purpose of considering locating his practice there.

Springer continued on his business for some two weeks, and upon his return to Cimarron, he found Stevens there, along with a company of cavalry soldiers from Fort Union. Stevens quietly informed Morley and others that the governor would be coming to Cimarron on the Saturday stagecoach to meet with those men—a dozen or so, including Clay Allison—who had signed the letter inviting him to visit Colfax County. Stevens's actual intention, as directed by the governor, was to effect the arrest of Allison for unspecified crimes. The wary men of Colfax did not show for the meeting, and neither did the governor. The incident would later assume greater importance, but for the time being it marked the end of efforts to influence the governor relative to the matter of the removal of the district court.[22]

As a sorry footnote to the affair, three of the black soldiers dispatched to Colfax County in support of Stevens's errand were shot dead in an altercation at the Saint James Hotel. The men reportedly had gone to the bar seeking liquor, in violation of orders given by their commanding officer. Frank Springer was present at the time and later recalled that when violence appeared imminent, he "stepped out the opposite door and retired from the scene as rapidly as my physical training permitted."[23] Accounts of the incident vary, but Springer characterized it as a "barroom row" brought on by bad whiskey. It is generally conceded that the soldiers were killed by David Crockett, nephew and namesake of the fallen hero of the Alamo, and his companion Gus Hefron. The matter was considered by a grand jury sitting in Taos, but no indictment was returned against Crockett or Hefron.

One additional source of irritation to Colfax County residents opposed to Governor Axtell and the Santa Fe Ring concerned the removal of the elected county sheriff, O.K. Chittenden, in favor of an appointee more favorable to the Ring. Axtell later insisted that Chittenden had resigned the office, but Springer obtained a copy of Axtell's letter informing Chittenden of his "removal from the office of Sheriff of Colfax County."[24] The removal was on grounds that the sheriff had failed to file a proper bond, as required by law. Chittenden's explanation was that "Judge Waldo refused to accept the bonds I gave and I refused to give any others."[25] To Springer and others, the administrative removal of an elected sheriff provided further evidence of the Ring's determination to manipulate political and legal affairs in order to maintain control in Colfax County.

To more serious grievances against the governor was added the charge that he was a bishop in the Church of Jesus Christ of Latter-Day Saints, and that his business in New Mexico included paving the way for a Mormon colony. Axtell at times vigorously denied the charge and at other times made fun of it, but the rumors were widely publicized in the New Mexico press, the governor chided as "El Obispo," the bishop. At one point Brigham Young tried to defuse the controversy, denying in a letter to Secretary of the Interior Schurz that Axtell was a bishop or even a baptized member of the church.[26]

Springer for the Defense

For a time other events had commanded public attention to the exclusion of the Tolby murder and its tragic aftermath. But in May 1876, a Taos grand jury indicted the Reverend O. P. McMains and fourteen others in the death of Cruz Vega.[27] McMains's friends and sympathizers responded by raising funds—some thirteen hundred dollars—for his defense. Frank Springer was retained to defend McMains and was later joined by Vermejo resident W. D. Lee, acting as co-counsel. Springer's practice in New Mexico had involved little in the way of criminal matters; suits for ejectment and defense of the Maxwell Company in contract disputes and other matters seem to have formed the bulk of his legal work.

Upon hearing of his indictment, McMains made himself available for arrest in Cimarron and was incarcerated in the adobe jail. Springer attempted to have him released on a writ of habeas corpus, but when bail was set at twenty thousand dollars, McMains was consigned to await trial in the dull confines of a jail cell. Various efforts were made to assist McMains during the months preceding his trial. Mary Tibbles McPherson, Morley's high-spirited mother-in-law, attempted to apply pressure through leaders in the Methodist Church. Following his marriage to Josephine Bishop on October 10, 1876, Springer and his new wife took a honeymoon trip to Springer's home state and points east. Springer took this opportunity to meet with William Fishback, an Indiana attorney with useful political connections, who also was McMains's brother-in-law.

A trial on the original indictment was aborted in the spring of 1877, but a new indictment was brought against McMains for the murder of Vega. Springer and Lee succeeded in obtaining a change of venue from Taos to Mora County, and trial was set for the August term of the district court. The trial was held at Mora on August 22–23. Judge Henry Waldo presided, while William Breeden, attorney general for the territory, handled the prosecution. Both men were presumed associates of the Santa Fe Ring, but the trial appears to have been remarkably free of political or personal animosity among the principals, and was so reported by the *Cimarron News and Press*.[28]

If the trial achieved nothing else, it did bring to light many sordid details concerning the murder of Cruz Vega. Prosecution witnesses included Isaiah Rinehart, a member of the coroner's jury that examined Vega at the scene of the lynching; William Low, the man McMains induced to lure Vega to a cornfield on the Rio Ponil near Cimarron; Irvin Lacy; and Manly Chase. Low was present during most of the confrontation between the lynch mob and Vega, and he detailed a sorry sequence of events involving unbridled violence on the part of several of the men present and a reluctance to interfere on the parts of others. Rinehart provided grisly details concerning the condition in which Vega's body was found and related evidence indicating that a violent struggle had taken place.

The trial concluded in a manner that satisfied almost no one. McMains was declared guilty "in the fifth degree," and a fine of three hundred dollars was specified. The *Santa Fe Daily New Mexican* found the verdict surprisingly lenient, but the *Colorado Chieftain,* a consistent supporter of McMains, took a jaundiced view of the proceedings, asking, "By the way, what sort of crime is murder in the 'fifth degree?' That is a dilution a little beyond anything we have ever heard before. But, we presume, it is like the justice dealt out to McMains, an article indigenous to the soil of New Mexico, and unknown to the ignorant outside barbarians."[29]

Predictably, the *Cimarron News and Press* took a position opposite that of its chief nemesis, the *New Mexican,* and in support of its former editor and printer: "We are informed that the universal sentiment of those who were present at the trial was that under the evidence and instructions of the court, it was impossible for the jury to do anything but acquit. The prejudice of race, however, defeated law and justice, the leading Mexicans on the jury favoring conviction on the ground that the grand jury of Taos County had found this man guilty of murder, and a Mora jury had no business to go contrary to this."[30]

The jury's verdict was rendered null and void when Judge Waldo accepted Frank Springer's motion for a new trial, based on the fact that the jury had not specified the crime of which McMains was found guilty. The succeeding events are perhaps best expressed in the words of Rev. Thomas Harwood, whose recollection across an interval of a dozen years was as follows: "Mr. McMains was present at the next term of court to meet the case. His attorneys asked for a change of venue from Mora County to Colfax, which was granted, and at the approaching session at Cimarron, in Colfax County, the case was thrown out of court by judge, now governor, L. Bradford Prince. Thus ended the long and vexed affair."[31]

By the time of McMains's trial at Mora, the Colfax County War—a political conflict accompanied by a series of inglorious murders—had become mostly a war of words. Clay Allison had quit the country, and David Crockett was dead, shot to death in October, 1876, in a confrontation with the sheriff. Times were more peaceful than they had been in a long while, but there remained a smoldering resentment against the powers in Santa Fe. In the early months of 1877, Governor Axtell had rejected a petition requesting restoration of the courts to Colfax County. The governor was authorized by law to return the courts to Cimarron after two terms had been held at Taos, but this he refused to do, to the consternation of Frank Springer and other residents of Colfax County.

Governor Axtell Boosted

In the spring of 1877, Mary Tibbles McPherson formalized charges against Gov. Samuel Axtell, addressing her complaint first to the U.S. secretary of the interior, Carl Schurz, and following up on May 3 with a letter to Pres. Rutherford B. Hayes. Several Colfax residents followed with affidavits echoing McPherson's sentiments relative to the governor.

Springer provided a sworn affidavit dated August 30, 1877, complaining about Axtell's actions concerning the removal of the district court from Colfax County. The major parties to the dispute, in Springer's mind, were the Santa Fe Ring and the anti-Ring citizens of Colfax County, but there were racial overtones as well. In a rhetorical excess that he may later have regretted, Springer made sensational allegations concerning the Hispanic residents of Taos County. His remarks would be ill considered by twenty-first-century standards, and even in his own time they were roundly criticized by his journalistic nemesis, the *Santa Fe New Mexican*. Springer asserted that residents of Colfax County could not obtain fair treatment in Taos owing to culture and language differences characterizing the dominant population groups in the two counties. He complained that juries thus far had been composed entirely of Taos County residents, few of whom could understand English. He elaborated as follows: "Affiant further states that among the inhabitants of Taos County, a large number belong to the religious sect called Penitentes, many of whom have been chosen upon the grand and petit juries, and that during the first term of said court after the removal thereof to Taos, these Penitentes held their public ceremonies in the town of Taos, consisting of the most horrible, disgusting, and barbarous, bloody exhibitions of self-torture, and that some of the jurors serving at that term, took part in these ceremonies, the grand jury adjourning one day to take part in and witness them."[32]

Axtell gave his detractors further cause to grumble when, in the summer of 1877, he visited Colfax County in the company of Sen. Stephen W. Dorsey of Arkansas. Axtell had hoped to interest Dorsey in a possible purchase of the Uña de Gato grant in the eastern portion of the county. Axtell again refused to meet with citizens in Cimarron, but reportedly took the opportunity to encourage residents in the eastern part of the county to "cut loose from the Cimarron crowd, and have a new county for themselves."[33] There was discussion of such a proposal in the succeeding session of the legislature, but legislation to that effect did not result.

A major victory was achieved when, in January 1878, the legislature moved to restore the district court to Colfax County. The event was reported in an extra edition of the *News and Press,* decorated with a rooster that appeared whenever the editors had something to crow about. Under the headline, "FROM THE CAPITAL/Courts Restored/Obnoxious Laws Repealed/The Legislature on Its Ear," the *News and Press* reported that an act restoring the courts to Colfax had passed both houses of the legislature on unanimous votes and that Axtell was expected to sign the bill.[34]

Springer must have derived a measure of satisfaction from this welcome news, but he was not done with the governor. In the spring, most likely in the first few days of April 1878, Springer came into possession of a document that caused him to come out of his seat with a vengeance. Familiarly known as the "Dear Ben" letter, the missive from Axtell to Benjamin Stevens was undated, but obviously from the tense days of March 1876, when Stevens was sent to Cimarron to arrest Clay Allison. In the letter, Springer found what he considered to be definite instructions from Axtell concerning the planned murder of Springer, Morley, H. M. Porter, Clay Allison, and others. The letter read as follows:

> Dear Ben—The second telegram delivered to you at Fort Union, directed to Cimarron, was intended to leak, but the operator here says he cannot raise the Cimarron office. If I was expected, our friends would probably be on hand, as the guard is only a government escort. I do not think your definite business is suspected. Wade informed Hatch that he had been ready all the time to assist you, but could not find that you wanted to do it. Hatch says their opinion is that you weakened and do not want to arrest the man. Have your men placed to arrest him and to kill all the men who resist you or stand with those who do resist you. Our man signed the invitation with others who were at that meeting, for me to visit Colfax—Porter, Morley, Springer, *et al.* Now, if they expect me Saturday they will be on hand. Send me letters by messenger, and do not hesitate at extreme measures. Your honor is at stake now, and a failure is fatal. If others resist or attempt murder, bring them also. Hatch is excited, and wishes, of course, to put all the blame on the civil officers. I am more anxious on our account than for any other reason. I clearly see that we have no friends in Colfax, and I have suspected all along that some of our pretended friends here were traitors.
>
> Yours, etc.,
>
> S. B. Axtell[35]

Springer wrote to Axtell on April 4, 1878, stating the facts as he understood them and requesting that the governor explain his actions. Axtell replied on April 7, denying any intent to injure Springer or others and stating that U.S. troops had been sent to Colfax County for the sole purpose of assisting the civil authorities in making arrests. Springer was not satisfied with the response and pursued an exchange of letters. The correspondence ended with Axtell's message of May 13, restating the purpose of Ben Stevens's mission and insisting that no violence had been intended against Springer "or any other peaceable citizen."[36] The *News and Press* was called into service as an organ of propaganda on behalf of Springer's position, publishing the "Dear Ben" letter on April 18, and again with a shrill commentary on May 23.

Still dissatisfied with Axtell's explanations, Springer wrote to Secretary Schurz, preferring charges against Axtell, requesting his removal as governor, and asking that the information be presented to President Hayes. Letters to Schurz and the president were accompanied by a sworn affidavit of ten pages in which Springer restated grievances concerning the removal of the courts from Colfax County and the governor's refusal to hear the people out. Based on the "Dear Ben" letter, Springer also accused the governor of conspiring to bring about the deaths of innocent citizens.[37]

President Hayes had already dispatched Frank Warner Angel to New Mexico to investigate charges against Axtell, but Springer's complaint no doubt added fuel to the fire. The troubles in Colfax County did not constitute the sum total of the governor's worries, nor were they necessarily the most severe. There was also major turbulence in Lincoln County, where two factions were warring and where John Tunstall, a rancher and a British subject, had been shot to death. While the Colfax County War had cooled considerably by this time, the troubles in Lincoln were still heating up, with no prospect for an early peace. Angel's commission directed him to investigate the governor's conduct relative to both of the troubled counties.

Frank Springer's father was still known in Republican circles and could on occasion be helpful to his son's cause. In the summer of 1878, Francis Springer traveled to Washington, carrying letters of introduction to Secretary Schurz and President Hayes. He reportedly discussed with the president the possibility of his removing Governor Axtell.[38] Judge Springer may also have discussed with the president likely candidates to succeed Axtell as governor.

By summer's end Angel had conducted numerous interviews and had addressed interrogatories to the governor. In the estimation of Victor Westphall, Frank Springer's deposition dated August 9, 1878, was the single most important statement influencing Angel's findings. Having previously indicated to Axtell that he had no charges to make against the governor, Angel did an abrupt about face after learning what Springer had to say, apparently even rethinking some of his findings on matters in Lincoln County. "Evidently Springer's account had changed Angel's mind all along the line," observes Westphall.[39]

Angel concluded his work in New Mexico by mid-August and left Santa Fe on August 17, 1878. Although the official report was not filed until October 3, President Hayes requested an informal report of Angel's findings soon after his return from New Mexico. On September 4, Secretary Schurz suspended Axtell—the suspension amounted to removal—and Gen. Lew Wallace, a Civil War hero and noted author, was appointed in his place. Recognizing that force might be needed to quell the disturbances in Lincoln County, George McCrary, secretary of war, equipped the new appointee with a letter of introduction to Gen. Edward Hatch, commanding officer for the District of New Mexico, requesting that Hatch provide such assistance as the governor might require.[40]

In Cimarron there was rejoicing. The celebratory rooster again appeared at the head of a special edition published by the *News and Press* on September 6. A barrage of headlines read:

> THE BISHOP BOOSTED!!
> AXTELL'S HEAD FALLS AT LAST.
> The Mills of the Gods Grind Slow, but They Grind Exceeding Fine.
> REJOICING IN CIMARRON.
> Fifty Guns Fired in Honor of the Event.
> General Lew. Wallace Appointed Governor.[41]

The paper ridiculed the deposed executive, reported the discharge of firearms and shooting of anvils in celebration, and concluded, "It will be now in order for the *New Mexican* and *Las Vegas Gazette* to wear the usual badge of mourning for 30 days."

The *New Mexican* jumped to Axtell's defense, publishing his rebuttal of the charges against him and chiding his chief detractor as "affidavit Springer" and "our affidavit man." On October 19, the paper lamented Axtell's departure, complimenting him for progressive and public-spirited government and hailing him as a person "held in the highest esteem for his ability, statesmanship and character" by the people of the territory.[42]

Meanwhile, the *News and Press* continued to pour it on the fallen governor, giving Frank Springer the last word in the paper's edition of October 3. Under the headline, "Who's a Liar?: Mr. Springer's Respects to Governor Axtell," Springer employed his best lawyerly rhetoric in dissecting Axtell's assertions. In an editorial note, the *News and Press* called attention to Axtell's and Springer's statements and urged readers to examine them carefully, adding, "We think the two will well pay perusal, as they bring out much that has not heretofore been made public, and we believe our readers will unanimously agree, that in this controversy the Ex-Governor is left 'too dead to skin.'"[43]

The two papers continued trading insults for several weeks, but the course of events was decided and the change in administration seemed to bring some closure to the years of controversy and bloodshed. With Axtell's dismissal, observes Norman Cleaveland, "The power of the Santa Fe Ring was broken, and although the Republican leaders made a comeback a few years later, they never again could control New Mexico to the extent they had for several years previously."[44]

Lew Wallace, age fifty-one, arrived in Santa Fe on the evening of September 30, 1878, and presented himself to Axtell on October 1. His overland journey from Trinidad was not the most comfortable, but it gave the new governor an opportunity to see the country up close.[45] En route Wallace stopped in Cimarron as the guest of Frank Springer. He was also guest of honor at a dinner and reception held in the home of Judge W. D. Lee. The *News and Press* noted the occasion and gave a hearty endorsement to the new governor, adding, "We feel confident that in securing for governor so courteous, able, and high toned a gentleman as Lew Wallace, the people of New Mexico are well entitled to congratulation."[46]

Wallace wrote to Secretary Schurz, advising of his arrival in Santa Fe and his qualification as governor. The new governor was pleased to report, "On my way from Trinidad, I stopped a day and a night in Cimarron, Colfax County, one of the localities about which a great deal has been said in connection with disorders. Without reference to the past, there certainly appeared to be a good feeling on the part of the citizens there, and a decided disposition to keep the peace."[47]

Feuding parties in Lincoln County and a young outlaw, William Bonney, otherwise known as Billy the Kid, would henceforth command a

Lew Wallace, territorial governor of New Mexico, 1878–81. Courtesy of Museum of New Mexico, Neg. 15295.

disproportionate share of the new governor's time and attention; Wallace assured the secretary that he would go to Lincoln and talk to the people immediately.

It seems reasonable to conclude, as did Angel, that Axtell's major sin, in Colfax as well as in Lincoln, was partisanship in circumstances that called for an evenhanded determination to learn the facts and act fairly and decisively. Angel found that there were enough "corruption, fraud, mismanagement, plots and murders" in the territory during Angel's tenure to warrant an immediate change in administration. Angel pronounced the governor "a complete tool in the hands of designing men."[48]

Bankruptcy and Reorganization

While the Colfax County War raged, the Maxwell Land Grant and Railway Company languished in a quagmire of debt, litigation, and uncertainty as to the status of its title.[49] Virtually none of the planned development of the grant's resources had taken place. Capital funds had been depleted, and by 1875 the company's securities were valued at a few pennies on the dollar. Squatters were numerous and their number was growing. The company was bankrupt and unable to pay its taxes or its officers' salaries, or to pay for other essential services. As a result, company officials could conclude agreements for the sale of land, but deeds were not being issued. In frustration, the company's own officers were contemplating lawsuits.

Nonpayment of taxes for 1874, 1875, and 1876 rendered the entire property vulnerable to a sale for taxes. A foreclosure sale was held by Colfax County sheriff Isaiah Rinehart on December 23, 1876, and the property was purchased for $16,479.46 by Melvin W. Mills. Mills was acting as agent for a group of investors or speculators headed by Thomas B. Catron; the interest purchased by Mills was subsequently conveyed to Catron. This transaction doubtless served to intensify tensions between Springer, Morley, and associates, on the one hand, and Catron, Mills, and other presumed Ring members on the other. The law provided for redemption of the property by its original owners if they could compensate the new purchaser, reimbursing Catron's group for the taxes owed and interest accrued. Control now passed to Dutch investors holding a majority of outstanding bonds.

Realizing the dire nature of the situation, the Amsterdam committee representing the grant's Dutch ownership retained Henry C. Thatcher and Charles E. Gast of Pueblo, Colorado, to investigate matters, formulate recommendations, and take certain actions on the committee's behalf. Gast looked into the matter and forwarded an initial report, confirming the grant's potential as a valuable property and detailing the financial and legal difficulties. He noted that the company had been without a president since the resignation of Stephen B. Elkins, but he complimented the efforts of Morley, Springer, and Harry Whigham as guardians of the company's interests. Of a lawsuit against the company on behalf of heirs of the former Mexican governor Manuel Armijo, Gast wrote, "Frank Springer, Esq. of Cimarron received a retainer from some source to defend the company. I have confidence in his management of the case, and I have instructed [him] to continue his services in the matter, and to advise Judge Thatcher and me of any steps that are taken."[50]

Most important, Gast noted that the property had been sold for taxes and he asked the committee's instructions concerning possible redemption of the estate. The Dutch were able to organize themselves sufficiently to pay Catron and associates the required sum of $20,961.85, and thus avoid a total disaster for the company.

Based on the recommendations of their American advisors, the Dutch might have been able to steer a course to legal and financial health had not unforeseen circumstances intervened. These included a lawsuit filed by the company's former president, John Collinson, and speculative maneuvers by an experienced promoter, Frank Remington Sherwin. These complications tied the company up in a web of suits, countersuits, and receivership actions, and prevented the more straightforward reorganization that the Dutch would surely have preferred.

Amid these difficulties, two rays of hope appeared. Pursuant to the finding of the U.S. Supreme Court in the case of *Tameling v. United States Freehold and Emigration Company,* in which the Court held that Congress had legally confirmed tracts exceeding eleven square leagues, Secretary of the Interior Schurz ordered the issuance of patents for comparable grants, including the Maxwell property. Although many financial and legal obstacles remained, this finding must have delighted the company's primary stakeholders.

Second, by the beginning of 1878, there was renewed, serious interest in extending a railroad into New Mexico and to the Maxwell property. Two companies, the Atchison, Topeka and Santa Fe and the Denver and Rio Grande, were exhibiting considerable interest in New Mexico, and were headed for a confrontation over control of Raton Pass. The AT&SF

prevailed, and by the end of the year track had been laid over the pass and into New Mexico.

By the time the Colfax County War had subsided, some favorable conditions were present for the Maxwell Company to right itself, proceed with efforts to quiet title to the grant, and pursue profitable development of the property. Alas, the company was in for yet another episode of dismal management and continued frustration. During this time, Frank Springer, local attorney for the Maxwell Land Grant and Railway Company, remained one of the few reliable constants, a highly capable and respected steward of the company's interests.

CHAPTER 4

Rails across the Territory

The restoration of courts to Colfax County and the departure of Samuel Axtell removed major irritants to anti-Ring citizens of New Mexico but did not bring an immediate or lasting peace. The troubles in Lincoln County continued; hostilities with native tribes erupted from time to time; and the question of title to the Maxwell grant remained in suspense—a continuing source of conflict for years to come. Nor was the Santa Fe Ring put out of business, but its activities, especially in Colfax County, were severely curtailed. The events of 1878, culminating in Axtell's removal and the appointment of Lew Wallace as governor, marked the end of a period of extraordinary turbulence and a resumption of efforts by enterprising newcomers to exploit the material resources of the vast and largely undeveloped territory.

Frank Springer again stood as a candidate for the territorial council in the fall election of 1878. He was a Republican by birth and political inclination but, perhaps because of the New Mexico Republican Party's strong association with the Ring, he ran on the "People's ticket." The *New Mexican* opposed his candidacy, deriding him as a "renegade," and reminding readers of derogatory statements about the Penitentes in Springer's affidavit of August, 1877.[1] The point was made more than once, in Spanish as well as English, and the paper especially urged that the district's Mora County voters give their attention to the matter. Springer, however, survived the attack and won election to the council, capturing substantial majorities in Colfax and Mora counties and defeating his opponent, Don Jesus Abreu. By the time the hostilities had subsided, Frank Springer's life had been altered materially, first, by his marriage; and second, by the long-awaited coming of the railroad.

Josephine Mary Bishop Springer. Photo by Furlong and Crispell Studio, Courtesy of CS Cattle Company.

Josephine Bishop was one of three sisters in a family that migrated from Kansas to New Mexico and took up residence in the Vermejo drainage. Clay Allison reportedly came courting one of the younger sisters until Mrs. Bishop saw fit to intervene, declaring him unwelcome and chasing him off the place with a broom.[2] Frank Springer, now twenty-eight years of age, pursued Josie Bishop with somewhat better results. On October 10, 1876, the two were married at Cimarron. Springer immediately took his bride east for a honeymoon and to meet his family in Iowa.

On January 12, 1878, Josephine gave birth to the couple's first child, Laura.[3] In anticipation of the event, Josie traveled to Columbus City, Iowa, where she was attended by her husband's family, and where the natal care presumably was superior to that found in an isolated frontier town. Following Laura's birth Springer wrote to his wife, imploring her to guard her health and avoid taking cold. "The baby will do well as long as you are all right," he advised, doubtless drawing more on a generous store of conventional wisdom than on any special expertise related to the matter. "If the roads get decent, you and Nellie had better ride out in the buggy occasionally," he added. "The open air will do you good."[4]

Steel Rails Coming

When Springer brought Josephine and the baby home to Cimarron, they traveled in the familiar manner, taking the train to the end of the line, then riding over rough roads to their final destination. This state of affairs was not long to continue, for by 1878, the railroads had awoken from the long slumber occasioned by a severe financial panic and the road builders were again pushing west and southwest. In some cases the competition for desirable routes and narrow passages through the Rocky Mountains was fierce. In southern Colorado, rival companies raced to secure critical rights-of-way, and armed guards were sometimes required to stand watch while construction laborers bent to their tasks.

One of the major players in this drama was Springer's good friend William R. Morley. Having left the active service of the Maxwell Land Grant and Railway Company, Morley was employed as a construction engineer for the Atchison, Topeka and Santa Fe. In this capacity he was involved in heroic efforts to secure possession of Raton Pass, and later of the "Grand Cañon of the Arkansas"—the Royal Gorge. In both instances Morley found himself in a battle with the Denver and Rio Grande. Tradition holds that Morley had earlier scouted the location on Raton Pass, ranging over the hills in the disguise of a sheepherder. His storied triumph came on the night of February 26, 1878, when he raced a crew from the D&RG to claim possession of the pass. The route was to be awarded to the first company to begin construction. Aware that the D&RG had diverted a crew to the area, Morley took a train to Trinidad and led a hastily organized party of local men in grading a few feet of roadbed by lantern light. At sunup the D&RG crew arrived and found, to their chagrin, that the Santa Fe men had preceded them.[5]

At dusk on December 6, 1878, the first rail was laid across the line into New Mexico. Morley was present and took part in the driving of ceremonial spikes. The *Trinidad Daily News* hailed the event as "Morley's Triumph" and congratulated the city's neighbors to the south: "The people of New Mexico can now say that they have a railroad into their Territory, although but few lengths of iron have been laid."[6] The *Cimarron News and Press* was no doubt thrilled with the arrival of the railroad in the territory, but with evident weariness, announced the occasion with the simple headline "At Last."[7]

In May, 1879, Josephine Springer was again in Iowa awaiting the birth of a child when she heard from her husband concerning the advance of the railroad. On his return from a business trip in Colorado, Springer had stopped to see Morley at the AT&SF construction site near Cañon City. Springer found it a busy place, with eight coaches a day carrying passengers to Leadville and other points, and rails being laid rapidly through the grand canyon of the Royal Gorge. Armed guards stood watch protecting the Santa Fe laborers from any possible mischief contemplated by agents of the rival Denver and Rio Grande.

More to the point, Springer could report that the Atchison, Topeka and Santa Fe was making rapid progress across the prairie southeast of Raton Pass. He was acutely aware of the significance of the transformation taking place. "When I returned to Pueblo I found the Directors' party of the AT&SF road with a train of five Pullman coaches, including a dining car. I was invited to join them, and we came down to Trinidad and the next morning over into New Mexico, to the end of the track, four miles beyond the Cimarron River. Moving down through Colfax County in such an elegant train made me feel that we had civilization at last, and that the age of the stage and the ox and mule train was past."[8]

According to Springer's later recollection, he was instrumental in bringing officials of the railroad and the Maxwell Company together as the

Santa Fe prepared to cross grant lands. "When the Santa Fe built into New Mexico money was not plentiful and business was light," he recalled. "Concessions or assistance tending to reduce the cost of construction were sought for, and regarded as of much importance."[9] Springer's business relationship with the Maxwell Company and his personal friendship with William B. Strong, general manager of the Santa Fe, enabled him to bring about "an extremely friendly understanding" between the two companies. As a result, the railroad received free right-of-way through the grant, station grounds at four points along the sixty-five mile stretch, good springwater for use in the railroad's locomotives, ties and bridge timbers for construction, one-half interest in the town sites of Springer and Raton, and one-half interest in ten thousand acres of coal-bearing lands.[10]

It soon became clear that Cimarron would not be on the railroad line, but Springer and other officials of the Maxwell Company were optimistic that an important town would be established on the grant. "It may be that the new town will be started in a month or so," Springer told Josie. "It has been named by the Railroad company, and I will bet a new pair of gloves you could not guess the name in six trials."[11]

Springer soon solved the riddle, telling his wife, "The new town was named 'Springer' by the railroad company, without my knowledge and against my wishes." Springer reported that the naming of the town had upset W. T. Thornton, court-appointed receiver for the Maxwell Land Grant and Railway Company, and that it was now a matter of contention but that he expected the flap to be settled in his favor.[12]

Springer began his work as a local attorney for the legal department of the AT&SF as soon as the railroad entered New Mexico.[13] His primary responsibility, especially through the months of engineering and construction of the route, was the acquisition of right-of-way. As Springer wrote to Josephine on June 4, he was preparing to leave on a journey of about one month. He was to travel west as far as the Rio Grande in a horse-drawn vehicle provided by the railroad company. "I rather dislike the trip at this time of year," he wrote, "but it is very important business and cannot be postponed. The railroad will be built on from Las Vegas 100 miles as soon as I get my work done, if I am successful."[14]

Citizens of Santa Fe became alarmed when they learned that railroad officials were planning to bypass the city and establish a station seventeen miles to the south. A meeting was called at Santa Fe in July 1879, for the purpose of discussing the matter. Howard Roberts Lamar has characterized the consultation as "virtually an 'estates general' of New Mexico."[15] Taking part in negotiations, in addition to Morley, Springer, and other railroad men, were Archbishop J. B. Lamy, Vicar-General Truchard, Territorial Secretary William G. Ritch, pioneer banker Jefferson Raynolds, Judge Henry Waldo, attorney William Breeden, W. H. Manderfield of the *New Mexican,* and other influentials.[16] It was finally agreed that the city would issue bonds to provide partial funding and a branch would be built off the main line, providing rail access to Santa Fe.

On July 26, 1879, Josie had her baby—a girl. Having been advised of the event, Springer sent congratulations to his wife. "I suppose of course the baby is pretty and looks like somebody," he wrote.[17] Claiming to have been overruled in the selection of a name for the couple's first child, Springer vowed to have a say in naming his new daughter. The baby was called "Helen."

Josephine stayed in Iowa through the fall and winter while railroad business kept Springer occupied and away from home much of the time. In addition, he was preparing to represent Colfax and Mora counties as a member of the council in the upcoming legislative session. From Santa Fe, Springer wrote telling Josie of his accommodations. His landlord was Archbishop Jean Baptiste Lamy, and the circumstances he described were not such as to evoke sympathy for a gentleman supposed to be "baching": "I am nicely fixed for the session, at last. I succeeded, after some trouble, in getting a good room in the house of the Archbishop, and now he has taken me in to board with him. Young Romero has a room next to me, and we live on the fat of the land—I guess the best table in Santa Fe. The very best of meats, and fruits, all the vegetables obtainable, good coffee and a bottle of wine at each meal. Breakfast at 9 and dinner at 5! So you may know I am happy."[18]

In other significant news, Springer reported that his brother Charles was going into the cattle business with him. Charles Springer had followed Frank to New Mexico and had settled at Cimarron. A few years after commencing their partnership, the brothers purchased the registered Herefords of Benjamin Herschey of Muscatine, Iowa, moved them to their father's property in Louisa County, and had the herd shipped to New Mexico by rail. With Charles providing on-site management, a concern first known as the Rocky Mountain Trading Company evolved into the Charles Springer Cattle Company, a diversified ranching operation.

Hereford cattle gathered at Judge Springer's home in Columbus Junction, Iowa, for shipment to New Mexico. Courtesy of CS Cattle Company.

A Place to Call Home

The coming of the railroad held profound implications for the social and economic future of New Mexico. It had another, more personal meaning for Frank and Josie Springer, in that it cast doubt on Cimarron as a suitable location. Having lost the railroad to the new town of Springer, Cimarron was certain to shrivel. Springer could read the signs clearly enough, and he began to contemplate his future. During the frequent times of separation from Josie, his letters reflected this concern.

His first thought was to build at the new town of Springer, but a few weeks later he was second-guessing the idea. "Everything is still in doubt as to our future headquarters," he told Josie. "There will be some strong inducements to stay at Cimarron, and in that case I will build here. I must say I would like it better than the new town, which I fear will not for some time be much more than a trading post for cattle men."[19] In another month, he had given up hope of building a home any time soon, "simply because there is too much uncertainty as to where the principal place is to be."[20]

In early 1882 Springer was still mulling the possibilities and tiring a bit of the perpetual uncertainty. "I would much rather be at Cimarron," he told Josie, "even if most everybody has left it. I don't know where another year will find us. I expect I will have to open an office somewhere else."[21] In April the Springer family welcomed a new member with the birth of another daughter, Eva. Apparently it was no longer considered hazardous for Josephine to remain in New Mexico for her "confinement," for the new baby was born at Cimarron.

On July 4, 1882, the county seat was formally moved to Springer, rendering Cimarron even more unsuitable for a thriving law practice. By August Springer was thinking of Las Vegas, New Mexico, as a possible location. "I think it would be a good deal pleasanter for you, and not so much out of the world," he theorized. "It is getting to be quite a city. I have some property here upon which we could build and have a good home."[22]

With the arrival of the railroad, Las Vegas had become one of the leading cities of the territory. Failing to extract sufficient concessions from the existing town of Las Vegas, the railroad builders had established their own town a mile to the east. Predictably, the old town reflected the territory's Spanish-Mexican heritage, while East Las Vegas was dominated by newcomers who brought with them the styles and customs with which they were familiar.

The railroad had built a sprawling resort hotel and bath house in nearby Gallinas Canyon and had built a spur off the main line to bring tourists to the spot. Moreover, Las Vegas was the commercial center for eastern New Mexico. Prominent mercantile firms—Gross, Blackwell and Company, Charles Ilfeld and Company, and Browne and Manzanares—headquartered there and traded all over the eastern half of the territory. The town's prosperity was evident in railroad shops, warehouses, and a growing residential district.

By April the matter was settled. "I think perhaps I have found something that may suit, in the way of a place to live," Springer reported. "There are two new houses of four rooms each, that are situated adjoining, and I can get them both, which would give us 8 rooms, altogether. The location is good, and there is a good back yard to turn the children loose in." Springer suggested sleeping in one house, cooking and eating in the other. "If you think this scheme will work," he concluded, "you might as well begin to think about moving, as we shall probably try to get moved about the last of May."[23]

The aspiration of building a comfortable home was deferred for the time being and the family made the move to Las Vegas on or about June 1, 1883. Within a few weeks Springer was advertising his relocated legal practice, with offices in the Dold Block, on the west side of the old plaza.

Another daughter, Ada, was born at Las Vegas on September 3. From the town of Springer, Frank wrote to say that he had learned of the birth from J. M. Cunningham, the attending physician and a family friend. "He did not give any particulars," Springer wrote, "but I learned otherwise, from friends on the train this evening, that the newcomer is No. 4 instead of No. 1." To dispel any hint of disappointment in having four daughters without a son, Springer assured his wife of his complete satisfaction concerning the outcome.[24]

Springer had not been present for the birth of any of his daughters, nor would he be home for the births of his sons over the next few years. Throughout their marriage, Springer apparently felt at liberty to give priority to business, civic, and avocational interests while Josephine reared the children and managed the household. Whatever the effect on Josie or the marriage relationship, Springer's seeming nonchalance with respect to family matters leaves the impression that he was a benevolent provider with minimal involvement in the day-to-day lives of his wife and children.

The Return of Samuel Axtell

It must have come as a rude shock to Springer to learn in early summer, 1882, that Pres. Chester A. Arthur had nominated as chief justice for New Mexico none other than Samuel B. Axtell, a man Springer assumed had been run out of the territory for good. The vacancy resulted from the resignation of L. Bradford Prince, who later served as governor. For Springer, an attorney whose work involved daily practice in the courts of the territory, there could hardly have been a worse nightmare. The prospect of Axtell's confirmation must have been all the more disturbing to him, realizing that the chief justice also sat as district judge for the First Judicial District, including Colfax County.

Upon learning of the nomination, Springer devoted himself to its defeat with the same passion he had demonstrated in the fight to unseat Axtell as governor. The appointment was subject to confirmation by the

U.S. Senate, and Springer was determined that it would not be approved if he could help it. He again called on his father, and Francis Springer hastened to the capital.

On June 18, 1882, Springer wrote to his father in Washington, enumerating Axtell's past sins in a five-page letter. There was ample evidence, Springer declared, "to show that such a man is not fit to be entrusted with any official position under the government, least of all the responsible position of Chief Justice of the very territory from which he was removed."[25] The letter was intended to furnish ammunition for the fight, and Francis Springer made such use of it, forwarding it to Illinois senator David Davis, a member of the Judiciary Committee. Francis Springer also advised Davis that a delegation was en route from New Mexico in hopes of testifying before the committee.[26]

The New Mexico delegation consisted of Frank Springer who, with his father, made the rounds in Washington for several days, working in opposition to the Axtell nomination. Upon his arrival in the city, Springer reported that his father was well and "busy as a beaver." Of chances for the success of his mission, he told Josie, "I guess we will soon have my old friend Axtell's scalp—or he will have ours, one or the other. I think we have him beaten, as the matter was sufficiently delayed on receipt of my telegrams, to enable me to get to work."[27]

Springer produced an eleven-page epistle for presentation to the Senate Judiciary Committee. He again recited Axtell's misdeeds and concluded, "After the experience of Mr. Axtell in two territories, from both of which he had to be removed, on account of the complaints of the people, is it prudent, is it right, to experiment further with him in one of the same territories, in an office whose maladministration would imperil every interest in that rapidly growing territory?"[28]

Springer also had allies at work back home. On June 28, W. D. Lee wrote from Las Vegas, advising of local efforts: "Yours of late date in regard to Axtell matter is at hand. Telegraphs, letters &c. have been forwarded from here. Petitions are being circulated and will be sent perhaps today. The people of this county are opposed to Axtell's confirmation."[29]

Although the *New Mexican* had changed hands, it remained on friendly terms with elements of the Santa Fe Ring and antagonistic to the Springer-led Cimarron faction. The *New Mexican* praised Axtell in its edition of July 2, and two days later reported, "A man named Springer is in Washington trying to defeat the nomination of Governor Axtell. Springer is a friend of the thugs and thieves of Colfax County."[30] The *New Mexican* also fired back in response to criticism of its position by the *News and Press,* and quoted from the *Denver Times:* "While the United States senate is investigating the charges made against the Hon. S. B. Axtell, it should also investigate those who make the charges."[31]

On July 4, Springer reported, "We are having a lively and interesting fight to beat Axtell's confirmation by the Senate. It looks as though we should be successful, but there's no telling."[32] On July 9 he was still in the capital, but sounding none so confident in the outcome. He advised Josie that the nomination was still pending and that there was some doubt as to the result. "I shall probably not stay to see it out," he wrote, "but shall let it take its course."[33]

Perhaps despairing of his mission's prospects, Springer left the capital and moved on to Boston. On July 15, he again wrote to Josie, conceding Axtell's probable confirmation and seeming to reconcile himself to that conclusion: "I left Washington with the Axtell matter in doubt, but in my opinion, with the chances rather in his favor. You need not borrow any trouble, or lose any sleep over the prospect of having him for Judge. I can get along with him, I have no doubt. New Mexico is different from what it was when he was there before, and I am different. I do not consider it in the power of any judge, however ill disposed, to do me any material injury, and I shall always be prepared to meet whatever situations present themselves in such manner as becomes my position and standing at the bar and as a citizen."[34]

Axtell's appointment had already been confirmed by the Senate on July 13 and reported in Santa Fe on July 14. The *New Mexican* exulted in the news and taunted its rival paper, declaring, "'Chief Justice Axtell' is a bitter pill for the Raton *News and Press*. Don't make a wry face, you've got to swallow it."[35] The paper hastened to report that Axtell's confirmation had come on the strength of unanimous votes in both the Judiciary Committee and the full Senate.

On July 18, a group of Santa Fe residents met for the purpose of planning a reception to welcome Axtell back to the territory. Perceiving that the event might give the appearance of an effort by elements of the Santa Fe Ring to gather in the new official, the *New Mexican* counseled against such a gesture.[36] Axtell arrived by rail on the evening of July 31, and reportedly was welcomed quietly by "a few old friends."[37]

Axtell served as chief justice until his resignation of May 11, 1885. There is no indication that further enmity of a public character occurred between the judge and Frank Springer, or that Springer failed to receive fair treatment in Axtell's court. Following his resignation as chief justice, Axtell remained in Santa Fe, where he pursued business interests and ran unsuccessfully for the office of probate judge. He died on August 6, 1891, while visiting relatives in New Jersey.[38]

The Continuing Saga of the Maxwell Grant

Pursuant to the March 1877, order of Secretary Schurz and following the completion of the authorized survey of the Maxwell Land Grant, a patent was issued on May 19, 1879, certifying the Maxwell Company's title to 1,714,764 acres of land. As surveyed, the property took in most of the western portion of Colfax County, extending into Colorado as far as Fisher's Peak and the Stonewall country. This event marked a major step in efforts to secure clear title to the property, but it was not, by a long way, the final act in the bitter conflict.

The company's opponents, chiefly settlers led by Rev. O. P. McMains, howled in protest and redoubled efforts to reverse the secretary's action. McMains hurled his wrath at proponents of the grant through his weekly newspaper, the *Raton Comet,* and railed against the grant men in spirited and sometimes incendiary orations preached to the faithful at meetings across the grant.

Meanwhile, the Maxwell Land Grant and Railway Company had been forced into receivership in June, 1878. An abundance of squatters and a paucity of paying leases and mining claims further complicated the company's position. Perhaps encouraged by the issuance of a patent and eager to take advantage of the company's weakened financial condition, Frank Remington Sherwin purchased major shares of the company's badly depleted securities and approached the Dutch bondholders concerning a possible foreclosure and reorganization. A lawsuit brought by John Collinson, the company's former president, introduced unanticipated complications, but the eventual result was a sheriff's auction held on March 22, 1880.

Sherwin and his associate Lucien Birdseye were the successful bidders, purchasing the entire estate, subject to first and second mortgages held by the Dutch bondholders. The concern was reorganized under laws of the Netherlands as the Maxwell Land Grant Company, with four Dutch and four American directors, with Sherwin as president. In consideration for their interest in the new company, Sherwin and Birdseye conveyed to the company the property rights they had purchased on March 22. To raise investment capital, the company issued a series of income bonds with a due date of May 31, 1920.

The *News and Press* expressed optimism, remarking of the new ownership on July 1, 1880, "We trust their promise to make the grant an exception to the dog in the manger policy that has disgraced the ownership of so many of the large tracts of land in New Mexico, will be redeemed, and that they will commence vigorously to develop their coal and mineral lands and make Colfax County, as it should be, the most prosperous of New Mexico."

Sherwin was somewhat more proficient as a promoter than as an executive. Almost immediately upon assuming management of the new company, he concluded an agreement for sale of coal to the Atchison, Topeka and Santa Fe Railroad that had the effect of binding the Maxwell Company's Raton Coal and Coke subsidiary to sell to the railroad at a loss. More eyebrows were raised in Amsterdam when the new president spent what appeared to the Dutch directors to be exorbitant sums of venture capital, with little to show. Neither were the directors pleased when Sherwin leased nearly 1.4 million acres to the newly created Maxwell Cattle Company in a transaction that smacked of blatant self-dealing.[39]

In October, 1880, the Dutch sent a representative, J. T. Nieuwenhuisen, to stand watch over the management and advise Sherwin of the Dutch directors' wishes. This action served only to incite conflict with Sherwin and with Frank Springer, in whom the Dutch directors had already invested proxies. An antagonistic relationship developed, as Sherwin refused to be bossed. Unable to effectively fulfill his intended function of imposing fiscal discipline on the Cimarron management, Nieuwenhuisen placed ineffectual advisories in the *News and Press* and sparred with the paper's editors in broadsides defending his own actions and excoriating both Sherwin and Frank Springer in the harshest terms.[40] Nieuwenhuisen returned to Holland a bitter and largely discredited envoy, and Sherwin continued, at least for the time being, on his merry way.

By the early months of 1883 it was clear that another management debacle was in progress and that the anticipated flowering of the grant's economic potential lay well in the future.

The End of an Era

A few months before his removal to Las Vegas, Springer lost a close and longtime friend when William R. Morley died of an accidental gunshot wound. Morley was in Mexico near the village of Aguas Calientes, locating a route for the Mexican Central Railroad, when the accident occurred on January 3, 1883. While riding in a carriage with other members of the party, Morley was struck in the aorta when a rifle was dislodged from its place and discharged.[41]

Many of Morley's friends, including Springer, H. M. Porter, Manly Chase, and Harry Whigham, were at the Las Vegas depot to meet the train carrying his body. Morley was laid to rest in the Odd Fellows Cemetery on the edge of the old town of Las Vegas. Just thirty-six years old at the time of his death, Morley left his beloved Ada and three children.

Morley's death and the move to Las Vegas marked the end of Springer's experience on the raw frontier of the nineteenth-century American West—an experience that involved railroad wars, the defeat of resisting Indian tribes, barroom gunfights, and violent political feuds. After settling in Las Vegas, Springer gave no further consideration to a permanent return to Iowa. Thirty-five years old as he took up residence in his adopted city, Springer settled into a career as a respected attorney in the busiest city of the growing territory, a mature and substantial citizen of East Las Vegas, New Mexico.

CHAPTER 5

U.S. v. Maxwell Land Grant Company

On April 19, 1887, the *Santa Fe Daily New Mexican* reported the conclusion of a long, highly publicized, and vigorously contested legal fight as follows: "Telegraphic advances reached Santa Fe late yesterday afternoon announcing that the Supreme Court of the United States had affirmed the decision of the great Maxwell Land Grant case, and proclaimed the grant, as made by the Mexican government by well-defined metes and bounds, valid in every particular. This news was not a surprise to anyone who knows enough to interpret the meaning of the treaty under which this vast property was guaranteed to Beaubien and Miranda, but for those cranks who maintain all grants to be frauds it proves a veritable black eye." Declared the *New Mexican,* "The history of this grant is the history of New Mexico."[1]

Having been at odds with the *New Mexican* and its political allies from the time of the Colfax County War and his brief tenure as editor of the rival *Cimarron News and Press,* Frank Springer was not accustomed to receive accolades from the Santa Fe paper. In the fate of the Maxwell Land Grant, however, Springer's interests converged with those of Thomas B. Catron and other influential Santa Fe citizens. The *New Mexican* was lavish in its praise, saying of Springer, "He has made a careful study of the subject for the past five years, and to him the people of New Mexico owe a debt of lasting gratitude for his faithful work in this connection, because, especially at this time when there is such general inquiry for lands in the southwest, the decision will go a long way toward forever settling the vexed and greatly exaggerated question of land titles in New Mexico." This case, more than any other, made Springer's reputation as an attorney and advanced his interests as a businessman, land owner, and developer.

To be sure, there were readers who would have disputed the notion that the Beaubien and Miranda Grant had been identified by "well-defined metes and bounds" or that the estate confirmed by Congress and upheld by the federal courts conformed to the intended limits of the original grant. Neither would the bitterly disappointed antigrant faction have agreed with the *New Mexican* that the Treaty of Guadalupe Hidalgo had been correctly interpreted and enforced with respect to the Maxwell Land Grant. There was, however, something to the paper's suggestion that there was much in the history of the Maxwell grant that could be generalized to the greater territory of New Mexico.

Said the *Raton Guard,* "The curse of New Mexico is the old Spanish and Mexican land grants that are spread all over the northern and central part of the territory." The problem was partially explained by the *Guard* as follows: "The worst feature about the grant system is probably the vast number of unconfirmed grants, and for which there is no sure means of ascertaining the true owners. Many of them are now controlled by the descendents of some old Spaniard or Mexican, who long ago received a grant of indefinite boundaries and locality, in consideration of services rendered by him to Spain or Mexico. These descendents have in several generations accumulated with a rapidity that is almost wonderful, and they are now scattered all over the territory. Some are rich and others were but lately peons and in the depth of poverty. To trace the original title through all the ramifications of Mexican genealogy, where many are gone, no one knows whither, and many are dead with no record of where or when they died, is almost an impossibility."[2]

In 1890, following a bitter and protracted litigation of the Maxwell title, Frank Springer's assessment was not much different. "The question is not whether we are in favor of grants or not," he wrote. "We all agree that the Territory would be far better off today if there had never been any."[3] The salient object, Springer declared, was and always had been a final disposition of contested titles, enabling all parties to escape the perpetual limbo of litigation and contentious administrative struggles and get on with business.

Springer approached the issue from the perspective of American law, which assumed a division of real property into public lands and private property held in fee simple. In New Mexico and on the Maxwell grant, two other traditions of land tenure were evident.[4] Practices inherited from Spain and Mexico involved different methods of conferring ownership and of measuring and designating holdings. The practice of conferring royal or executive grants for frontier defense and other purposes was well established, and in some communities common lands had been designated for the shared use of residents. Private ownership of land was a notion even more alien to the Utes and Jicarilla Apaches for whom northern New Mexico was a traditional homeland.

Complicating matters further was confusion over the status of lands claimed by the Maxwell Company. Given the radically differing assumptions and expectations of parties to the dispute, the determination of the Maxwell Land Grant controversy was bound to be traumatic for some of those involved.

The Beaubien and Miranda Grant

The land grant quandary was the legacy of a dubious war fought with Mexico when the notion of Manifest Destiny was at its zenith.[5] The war was settled with the Treaty of Guadalupe Hidalgo, signed by both parties and ratified by Congress in 1848. By this treaty, a vast territory, including all or parts of the present states of New Mexico, Arizona, California, Colorado, Wyoming, Utah, and Nevada, was acquired by the United States. An issue of great importance to the vanquished Mexicans was that of title to land grants made by Spain and Mexico. Under a protocol adopted by both parties, grants found to be valid under Spanish and Mexican law were to be considered valid under American law.

Six years passed before action was taken to deal with land claims in New Mexico. In 1854 a surveyor general was appointed, with instructions to examine all private land claims presented for consideration and forward to Congress his recommendations concerning confirmation or denial. The task set before the surveyor general was monumental and his resources meager; his cause was hopeless from the beginning, but the new appointee, William Pelham, went forth to execute the duties of the office.

Among the first claims to be considered by the new surveyor general was a grant made jointly to two citizens of Taos, Carlos Beaubien and Guadalupe Miranda. First granted by the Mexican governor Manuel Armijo in 1841, the estate was conveyed to the claimants in 1843 in a procedure prescribed by Mexican law and custom. This duty was performed by Cornelio Vigil, justice of the peace at Taos, who traveled with the claimants over the estate described in a rough sketch, or *diseño,* and established markers designating the corners of the property and defining its boundaries. The amount of land so conveyed was unknown and unspecified.

Between 1858 and 1867 Beaubien's son-in-law, Lucien Bonaparte Maxwell, acquired the property through purchase from Beaubien's heirs, and from Guadalupe Miranda, who had returned to Mexico following the American occupation of New Mexico. A native of Illinois, Maxwell had accompanied John C. Frémont on three of Frémont's western expeditions, and in so doing, had made a friend and business associate in Christopher "Kit" Carson.[6] Maxwell had engaged in ranching on the grant as early as 1847 and by 1848 had established a settlement at Rayado. Ten years later, he had established his headquarters at the present-day village of Cimarron. Following Maxwell's acquisition of the vaguely defined Beaubien and Miranda Grant, common usage referred to the property as the "Maxwell Land Grant."

Upon learning of the appointment of a surveyor general for New Mexico, and of his charge to recommend confirmation of grants found to be valid under Mexican law, Beaubien engaged an attorney in Santa Fe and began the process of petitioning for favorable action on the claim. By the time Maxwell gained practical control of the property, this process was well under way. Pelham found the Beaubien and Miranda claim to be in order, and the title to the property was confirmed by act of Congress on June 21, 1860. As yet, the extent of the Beaubien and Miranda or Maxwell Land Grant had not been determined. Opinions and estimates ranged from approximately 97,000 acres, representing grants of eleven square leagues each to Carlos Beaubien and Guadalupe Miranda, to 2 million acres, a figure based on the most liberal interpretation of original grant documents.

Pursuant to an act of Congress passed March 8, 1869, the surveyor general ordered a survey of the Maxwell grant. Under contract with the surveyor general, William W. Griffin of Santa Fe commenced the survey, relying on maps, sketches, testimony, and other materials filed in connection with the original claim. However, the survey was suspended on December 31, 1869, on order of Secretary of the Interior Jacob Cox, who had learned of the unusual size of the tract to be surveyed. The secretary believed that the Beaubien and Miranda claim could not legally exceed the common limit of eleven square leagues for each claimant.

Cox's ruling came at a critical time, just as Maxwell was contemplating a sale of the property to a group headed by Colorado senator Jerome B. Chaffee. The prospective buyers arranged with Griffin to complete the work he had started, and the resulting plats were filed with the General Land Office. The new owners of the property, the Maxwell Land Grant and Railway Company, approached Cox's successor, Columbus Delano, to plead for a reversal of the former secretary's decision. Delano expressed sympathy with the company's argument, but declined to substitute his judgment for that of Secretary Cox. The owners of the Maxwell grant were welcome to identify a tract equal to eleven square leagues for each of the original claimants—a parcel of approximately 97,000 acres. This was not what the investors had anticipated in purchasing the grant from Lucien Maxwell and his wife, Maria de la Luz Beaubien Maxwell.

Refusing to accept a patent for a mere fraction of the original claim, company officials requested a return of funds deposited with the General Land Office for the purpose of securing an official survey of the property. Willis Drummond, commissioner of the General Land Office, struck back with an order that further diminished the company's position. On January 28, 1874, he wrote to James K. Proudfit, surveyor general at Santa Fe, instructing him as follows: "As the claimants have withdrawn their whole deposit and refused to comply with the terms of the decision, you will regard and treat the lands claimed by them as public lands, and extend the public surveys over the same."[7]

Had the struggle been solely a private matter between the government and the Maxwell Company, the decision of Secretary Cox and the order of Commissioner Drummond would have been but a couple of blows in a protracted legal bout, the outcome of which was still far off. However, given the large number of settlers threatened with the loss of homes and ranches, these rulings had the effect of perpetuating a bitter, occasionally violent, and very public struggle over ownership of lands claimed by the company.

Springer's former ally and client Oscar P. McMains emerged to lead the dissident settlers. His object was to nullify or restrict the Maxwell claim, and it was a cause to which he devoted the last two decades of his life. McMains was at times a genuine hero and at times a doubtful one to the squatters, but he remained a dependable thorn in the side of the grant's management. His opposition to the Maxwell Land Grant and its various owners placed him squarely at odds with the company's legal counsel, Frank Springer.

While the Maxwell Land Grant and Railway Company struggled financially, a significant breakthrough came with the Supreme Court's 1877 decision in the *Tameling* case. In dealing with John G. Tameling's claim to the million-acre Sangre de Cristo grant, the Court had determined that Congress could, if it wished, approve a claim larger than the conventional standard of eleven square leagues per grantee. Congress could create, in effect, a grant de novo, or a new grant. If sustained, this view rendered moot the question of whether a grant conformed to the conditions prescribed under Spanish or Mexican law.

J. A. Williamson, commissioner of the General Land Office in Washington, was quick to note the similarity between the Sangre de Cristo claim and that of the Maxwell Land Grant. He called the matter to the attention of Secretary of the Interior Carl Shurz and requested the secretary's guidance. Williamson understood Schurz's reply to direct that he reopen the Maxwell case and order a new survey, preparatory to the likely issuance of a patent. As the controversy wore on, the monumental importance of the ruling on the Sangre de Cristo claim became apparent. In the view of O. P. McMains's biographer, Morris Taylor, "The Tameling decision was the impregnable bastion from which assaults against the Maxwell Grant could be and were repulsed; it was the immovable obstacle that McMains and the antigranters could neither eliminate nor circumvent."[8]

The new survey was executed by John Elkins, a brother of Stephen Elkins, and Robert Marmon. The survey resulted in a minor exclusion of land relative to the survey completed in 1870 by William Griffin. Proponents no longer talked in round numbers, of a grant approaching 2 million acres, but at 1,714,764.94 acres, the property as surveyed was still substantial. In the minds of McMains and his followers, it was fraudulent and outrageous. When a patent was issued based on the Tameling decision, the antigrant forces aired their grievances in the *Comet,* an antigrant paper published by McMains. Backed by a government patent, the Maxwell Company instituted lawsuits against settlers living on the grant without satisfactory titles. Meanwhile, McMains preached defiance and refused to purchase the parcel of land upon which he was living near Raton.

With the support of the antigrant settlers, McMains traveled to Washington in June 1881, to meet with officials and request that a lawsuit be brought to vacate the patent of the Maxwell Land Grant Company. The attorney general declined to pursue such litigation, and Frank Springer was so advised by letter from the Justice Department.[9] Secretary Kirkwood confirmed this information on behalf of the Interior Department in a letter dated December 30, 1881.[10]

THE PATENT SUIT

This would seem to have settled the matter, but the Maxwell controversy did not die easily. In 1882 Benjamin H. Brewster was appointed attorney general by Pres. Rutherford Hayes. McMains continued to plead the settlers' cause, apparently with good effect, for in March 1882, Brewster reversed the decision of his predecessor and ordered a suit to seek cancellation of the Maxwell patent. Exultant, the *Raton Guard* rallied the settlers, labeling the claims of the Maxwell Company "preposterous and fraudulent," and extolling McMains as "the brave man who has made so noble a fight."[11]

Loath to let McMains enjoy his victory, Maxwell Company officials served ejectment notices on McMains and several of his neighbors. McMains was in Washington when the notice was served, but hay and other goods were attached in his absence. In pursuit of the order against McMains, the sheriff scheduled a sale of his livestock. The event drew a large crowd of antigrant settlers; when bids were called for, emboldened supporters of McMains let it be known that his livestock would not be sold. No bid was proffered, and the sheriff departed empty-handed.

In July 1882, McMains purchased the *Guard,* renaming his paper the *Raton Comet*—he had suspended publication of the original *Comet* in mid-1881—and announcing as its motto, "Open war against secret fraud." He immediately declared war on his rivals, promising in the first issue that the paper would "go for the diseased and unwieldy corporeity of the Maxwell Grant hot and heavy."[12]

The suit to vacate the Maxwell patent was filed in the federal circuit court in Colorado. This strategy met with the approval of antigrant supporters, who believed that the grant was most vulnerable in the matter of its northern boundary. There had long been rumors of men hired to move one of the Colorado markers several miles north, and many believed the survey to have been fraudulent. McMains was convinced that the Colorado suit would produce the desired effect of putting the Maxwell Company in its place.[13]

If Springer needed any special motivation for his work on behalf of the Maxwell Land Grant Company, it was to be found in the original bill of complaint filed in the federal circuit court. In advancing its claim of fraud, the government did not confine its accusations to the Maxwell Company and the railroads that had come into possession of grant lands, but named the company's principal officers at the time of the alleged fraudulent act—Stephen B. Elkins, president; Harry Whigham, secretary and treasurer; and Frank Springer, attorney. According to the complaint, these men "wrongfully and fraudulently" contrived and conspired "to cheat and defraud the United States out of its said lands."[14] The government's bill suggested that there had been collusion between the company's officers and the surveyors, resulting in the inclusion of a large tract of land not contemplated in the grant to Beaubien and Miranda. Springer was not inclined to suffer such an attack on his integrity, and he addressed the matter when the opportunity presented.

In late summer 1883, two attorneys representing the Justice Department traveled to the Southwest to depose witnesses and otherwise build the government's case. On August 3, 1883, the *Comet* reported that the men were to examine records in the Santa Fe office of the surveyor general and that they would "take the testimony of parties who are well informed as to the ancient land marks of this country, upon which the boundaries of the grant are established."

Upon completion of the investigation, the government filed an amended bill of complaint and Springer and Gast filed a demurer in response. Curiously, the government's amended bill did not contain the charges made previously against Elkins, Springer, and Whigham. Springer and Gast requested an immediate decision in favor of the Maxwell Land Grant Company on grounds that the validity of a grant larger than eleven square leagues per claimant, already confirmed by Congress, was assured by virtue of the *Tameling* decision, and that as innocent purchasers, the Maxwell Company should not be called to account for possible error in a survey commissioned by the government preparatory to the issuance of a patent.

While the two sides were preparing to do battle in court, another skirmish was taking place in the local press. Springer and O. P. McMains locked horns in the *Raton Comet* and kept up a rhetorical feud for many months. In early 1884, following the government's amended filing and the Maxwell Company's response, Frank Springer took exception to a published account of O. P. McMains's remarks at an antigrant rally at Raton. McMains had accused the grant's attorneys of "setting up" the *Tameling* case to serve their own ends. He further asserted that, in allowing the filing of an amended bill, Judge Moses Hallett had ruled that *Tameling* had no bearing on the Maxwell dispute. Believing McMains's statement to be a gross abuse of the truth, Springer wrote to the *Raton Comet* to correct the record. Said Springer in closing, "I have no objection to Mr. McMains having credit for all the 'victories' that have been gained thus far, but I do object to his imposing upon the people of Colfax County accounts of decisions of the court that have never been made."[15] McMains fired back, reiterating the charge and alluding to possible collusion between Springer and Judge J. A. Bentley, the government attorney in whom McMains was rapidly losing confidence.[16]

The two men traded accusations in the *Comet* over the next month, McMains making more explicit his suspicion of a conspiracy between the attorneys for the Maxwell Company and the attorneys representing the government."[17] Springer's final word on the matter was brief, emphasizing the essential point that the judge had not ruled the *Tameling* decision inapplicable to the Maxwell case. "Whatever questions exist between Mr. McMains and Judge Bentley *they* must settle," he wrote. "They do not interest me in the least. Mr. McMains attempted to humbug the people, and was exposed. He is naturally furious over it, and tries to cover his tracks by abusing me. I can stand it."[18]

The Brewer Decisions

By early summer the conflict had shifted from the newspaper to the Denver courtroom of David Josiah Brewer, judge of the Federal Circuit Court for the State of Colorado. Much was foreshadowed when the judge ruled on the Maxwell Company's demurrer in July 1884. Of greatest significance was Brewer's finding concerning the applicability of the eleven-league limit for individual grantees. "I think the case of *Tameling v. Freehold Co.* 93 U.S. 644, effectually disposes of this question," the judge wrote. "That case held that the confirmation by an act of congress was equivalent to a grant *de novo,* and I see no substantial difference between that case and this."[19] Brewer did, however, rule that the Maxwell Company would have to defend the merits of its claim concerning the disputed Colorado portion of the grant and the boundaries upon which that claim was based.

Oral arguments would not be heard for more than a year, but Springer and McMains kept the pot boiling, taking up their war of words again in the *Comet.* The paper was no longer owned by McMains but it remained sympathetic to his cause, and usually to him. McMains led off this volley, returning to the 1869 ruling of Secretary Cox, which McMains still held to be the authoritative statement on the status of the Maxwell grant. McMains insisted that the bulk of the Maxwell claim was public domain by virtue of the Cox decision.[20]

On October 10, 1884, Springer directed a long epistle to "the Settlers Upon the Maxwell Grant." Confident that Judge Brewer's application of *Tameling* would sustain the Maxwell claim, he sought to assure antigrant readers that much less was at stake than they imagined. "Not a single man of you living south of the Raton Divide will gain one foot of land, if the government wins the patent suit," he asserted, "and as to those who live north of the Raton Divide I say that you are far more likely to be struck by lightning than you are to be benefited by the patent suit."[21]

In June 1885, a new player entered the controversy in the person of George Washington Julian. Over a long career in public life Julian had been a Whig, a Free-Soiler, and a Republican. Having failed to secure renomination as a congressman from Indiana in 1870, Julian bolted the Republican Party and joined the Democrats.

At the age of sixty-eight, Julian was appointed surveyor general for New Mexico by Pres. Grover Cleveland in an apparent effort to reform a land system characterized by boondoggle and manipulated by the Santa Fe Ring.[22] Julian embraced the task, reviewing some thirty-five claims, recommending rejection of some and reduction of others. The administrative and judicial disposition of the Maxwell claim was too far along for Julian's efforts to have any substantial effect, but his public pronouncements may well have encouraged further resistance on the part of defiant settlers.

Preparatory to oral argument of the case before the Colorado circuit court, attorneys for both sides took testimony from witnesses at Las Vegas in May 1885. J. A. Bentley questioned witnesses on behalf of the Justice Department, while Springer represented the Maxwell Company. O. P. McMains was among those called by the government. Jesus Silva, a member of the party that placed the original cornerstones in 1843, and Richens Lacy "Uncle Dick" Wootton, proprietor of a toll road over Raton Pass prior to the coming of the railroad, provided testimony favorable to Springer's case. Wootton had witnessed the building of a crucial boundary marker and could testify that the line surveyed by Elkins and Marmon conformed to original boundaries of the grant as established by the Mexican authorities.

Springer also chose to testify. He detailed his role in providing information to John Elkins preparatory to execution of the disputed survey. He recalled having directed Elkins to Jesus Silva as a knowledgeable witness, and he admitted having discussed with Elkins the company's claim relative to the northeastern boundary. He stated that no effort had been made to induce the contractors to provide a survey favorable to the company's interests and

Jesus Silva and Richens Lacy "Uncle Dick" Wootton. This 1885 photograph was likely made in Las Vegas, New Mexico, in May, when depositions were taken in the Maxwell title suit. Courtesy of Museum of New Mexico, Neg. 13112.

cited, as an example of their independence, the exclusion of a tract claimed by the company in the vicinity of the Uña de Gato Creek. Finally, Springer disclosed his interest in testifying:

> I make the foregoing statement for the reason that in the bill of complaint originally filed in this cause, and signed by the Attorney General of the United States, I have been personally charged with conspiracy with the said Elkins and Marmon to defraud the Government, by inducing them to make a false survey; a charge which I here declare to be false in fact and baseless in foundation, and a wholly unwarranted attack upon the character of a private citizen; I now invite the fullest cross-examination, not only as to the facts which I have stated, but as to any other matters involved in this suit, or mentioned in the pleadings, which may be within my knowledge, and for that purpose I waive every privilege of my position as counsel.[23]

J. A. Bentley, counsel for the government, asked two questions intended to clarify Springer's testimony, but did not challenge the witness further.

A full hearing on the Colorado patent suit came in the fall of 1885. The judge's decision was not announced until early 1886, but the *Denver Tribune* took pleasure in furnishing its readers with detailed accounts of the opposing arguments. Springer's defense followed the expected course of citing the *Tameling* decision, defending the validity of the Elkins and Marmon survey, and repelling charges of fraud. "The argument for the Government in this case consists principally of adjectives," Springer told the *Tribune*. "It abounds in allusions to 'falsifications,' 'forgery,' 'robbery,' 'brazen and bare-faced falsehoods,' etc. Probably no case ever better illustrated how much easier it is to charge 'fraud' than it is to establish its existence by proof."[24] Springer returned to New Mexico confident that the outcome would favor his client.

Judge Brewer issued his opinion on January 25, 1886. He made quick work of the first question, ruling that the *Tameling* decision was both applicable to the Maxwell case and binding upon the court. Concerning the second issue, Brewer ruled that the survey made by Elkins and Marmon was as accurate as could be expected, given the vague nature of the original grant. As to the last of the Justice Department's contentions, that Maxwell, subsequent owners of the grant, or both had contrived to defraud the government, Brewer sided with the defenders of the Maxwell claim: "So far as the allegations of fraud and intentional wrong on the part of the surveyors or the claimants, there is not a syllable of testimony worthy of the slightest consideration. The surveyors had no interest in the matter. They followed the instructions of the land department. They acted on the only testimony presented to them, or within their knowledge. They had no dealings with the parties interested in the grant, save as the latter protested against their action in excluding the Uña de Gato tract from the survey. If ever there was an official act in which no improper conduct was shown on the part of the officials it is this survey. The same may be said of the conduct of the claimants."[25]

Brewer's decision was reported widely in the national press. Back home Springer received the accolades of friendly papers, while suffering the expected slings and arrows from the antigrant press. He also received the personal congratulations of friends and colleagues, including Stephen B. Elkins, who by now was well situated as the son-in-law of the industrial tycoon and former U.S. senator H. G. Davis of West Virginia. "You have done well," Elkins wrote. "The opinion is a full answer to the lies and calumny that have floated around through the press for years. I am curious to know what McMains will say now."[26]

McMains had already expressed himself on the matter, addressing an antigrant crowd in Raton four days after Judge Brewer's ruling. McMains was just in from Washington, where he had been pleading his case before President Cleveland, Secretary of the Interior Lucius Lamar, territorial delegate Antonio Joseph, and a congressional committee on private land claims.

On the Maxwell grant the Brewer decision served to fuel anger and heighten anxiety on the part of settlers who stood to lose their homes. Having won the latest legal battle, the Maxwell Company had begun efforts to eject those against whom judgments had already been issued. The company's intentions were unclear, and it was rumored that the company was attempting to have troops ordered to the area.[27] Harry Whigham, court-appointed receiver for the Maxwell Land Grant Company, pushed the company's agenda and took most of the heat for actions judged precipitous by the antigrant faction. Troops were not sent to Colfax County, but relations between settlers and company agents remained tense.

In the March 26 edition of the *Comet,* Whigham had a long letter, conciliatory in its tone, attempting to dispel misunderstanding and offering to settle with all who claimed lands on the Maxwell grant.[28] The epistle was well written, eloquent in places, and appealed to reason on the part of the reader. The letter drew a response in the *Raton Weekly Independent,* from a writer identified only as "Fair Play," opening with the skeptical comment, "I see a letter published in the *Comet* of last week written by the Hon. Frank Springer whistled through Harry Whigham, secretary and receiver of the

Maxwell grant company." Said the writer, "The gist of the letter is to prove to those of us who know little or nothing about the steal that the managers of the fraud are high-toned gentlemen and are slopping over with sympathy for the poor settlers. We spurn the proffered sympathy and spit defiance at the land robbers and their hirelings."[29]

Following a meeting between Whigham and representatives of the settlers, the *Comet* reported that reason had prevailed and that no further attempts would be made to eject settlers until the U.S. Supreme Court had issued a final decision.[30] Unwilling to let the matter rest, O. P. McMains issued a new broadside titled "Conspiracy to Defraud." McMains repeated a familiar litany of charges, stated nine points of fact according to him, and excoriated the government for allowing public lands to be plundered by "a hypocritical, lying, thieving, bribe-polluting foreign corporation."[31]

Shortly after the issue of his broadside, McMains received an unkind cut from the *Raton Comet,* which he had previously owned and which had supported the antigrant cause. Declared the paper, "The *Comet* is fully satisfied that from now on the courts and not the agitators will guide the destinies of the citizens of Colfax County, and while it has and ever will advise the settler to go as far as the law allows in asserting his rights, it can not in justice advise those who have their all at stake to be guided by a man whose life long teachings lead to anarchy and mob law, and who has nothing to lose and everything to gain by disorder and tumult."[32] While acknowledging McMains's past services, the *Comet* had concluded that he was no longer an effective spokesmen for the antigrant cause.

Before the High Court

Having invested heavily in the case against the Maxwell Company, the Justice Department could hardly let the matter drop without pursuing a final hearing before the U.S. Supreme Court. Though many of the antigrant settlers had lost hope that judicial proceedings would grant them any relief, McMains and others remained optimistic.

An appeal was filed in the 1886 October term, and Springer joined counsel for the government in requesting that the cause be advanced for an early hearing. The government's brief was succinct, but Springer and Charles Gast plied the justices with a printed brief of some 105 pages, complete with a recitation of the history of the conflict and maps showing the Maxwell grant as located in the old Mexican *diseño* and in more recent surveys.

The case was argued in March 1887. William Maury, the assistant attorney general, and special counsel J. A. Bentley presented the government's case, while Springer and Gast answered for the Maxwell Company. The substantive questions were much the same as those Judge Brewer had dealt with, centering on the legality of the grant, the accuracy of the north boundary, and the allegation of fraud.

On the morning of the third day, the defense was called to present its case and Springer rose to represent the Maxwell Land Grant Company, opening his remarks as follows: "Neither one of the counsel who have preceded me has given an orderly and systematic statement of the case, as was incumbent upon them in the opening. I shall therefore proceed to supply what they have omitted. The court knows the law better than we do, but it cannot easily learn the facts contained in a thousand pages of record, unless counsel, whose business it is to be familiar with them, shall give the court the benefit of their knowledge; and if it consumes every moment of the time that may be allowed me, I intend that when I get through this argument the court shall know *what* it has to decide, whether I may contribute anything as to how it should decide or not."[33]

Springer proceeded with a history of the Maxwell claim, touching on each significant event up to and including the decision of Judge Brewer in the circuit court. Having given over much of his allotted time to a historical recapitulation of the case, Springer was granted additional time on March 11. In closing, he noted that the territory in question had been under U.S. sovereignty for some forty years and that the active controversy surrounding the Maxwell claim had consumed more than twenty-five years. "There ought to be some time in the history of a title when it will be safe to deal with it," he declared. "There ought to be some time, during the life of a generation, in which a title, for the due acknowledgment of which the sacred honor of the nation is pledged, shall become settled so far as the action of the government is concerned."[34]

Springer lamented the damage suffered by his client from the government's failure to resolve the title dispute, but noted that the effort expended in litigation would not be in vain if it should result in "an effectual declaration by this court of the principles which shall hereafter guide the officials of the United States as to the hundreds of other titles in the territory from which I come."[35]

Springer's mentor and former law partner, Henry Strong, was present to hear the argument before the Supreme Court. Immediately following the proceedings Strong wrote to his old friend Judge Francis Springer of Iowa. "Your paternal heart would have swelled with pride had you heard Frank for the past two days in the Supreme Court of the United States," he wrote, telling the elder Springer, "I have never seen counsel appear to better advantage. I have never heard an abler, clearer, more convincing argument in support of the legal positions taken. The court were interested, instructed, and most flatteringly attentive. Frank's manner and delivery are also admirable. I have little doubt that he will win his case—in fact I have no doubt of it."[36]

Strong's confidence in the outcome was based on more than his considerable knowledge of law and the judicial system, for he had left the Court's chamber in company with Justice Samuel Miller, a longtime Iowa attorney and a friend of both Strong and Francis Springer. Strong reported Justice Miller "unreserved in his expressions of admiration of Frank's argument both in matter and manner."[37]

The Court announced its decision on April 18, 1887, Justice Miller writing the opinion on behalf of a unanimous Court. He might have made short work of the decision, since it essentially affirmed that of the Denver circuit court, but he went over each of the contentions of the government, citing precedents and giving the reasoning of the Court. The Court declined to overturn the action of Congress in confirming the Maxwell claim, rejected the allegations of fraud, and affirmed the decree of the circuit court.

Springer was again deluged with editorial praise and the congratulations of business and professional associates. The *New Mexican* was jubilant, calling the Court's decision "the most inspiriting event that has ever occurred in the history of New Mexico."[38] After the decision had been announced, Justice Miller wrote to Francis Springer, enclosing a copy of the opinion and applauding Frank Springer's performance before the Court. "It was not only the force of the reasoning which pleased me," he wrote, "but the easy and graceful manner, the absence of rant and overstrained voice, and the orderly arrangement of what he had to say, which is so seldom seen in our Western young men."[39]

O. P. McMains, on the other hand, caught an early train for Washington with evident intent of containing the damage, or perhaps doing some himself. He successfully urged officials at the General Land Office to press for a rehearing of the case. He also sought a new lawsuit against the Maxwell Company for the purpose of challenging the patent in New Mexico, and he requested and received an audience with President Cleveland.

Giving vent to his frustration with the Supreme Court, McMains unleashed an angry tirade in which he called the justices of the high court liars and corruptionists and expressed his deep sense of shame in them. The diatribe, styled "An Open letter to the Justices of the Supreme Court of the United States," was printed and broadcast over the city of Washington, and was published in the *Raton Daily Range* on May 20.[40] Springer furnished a clipping from the *Range* to Justice Miller.

If McMains's screed was calculated to move the Court to favorable action, it missed the mark. The Court took only fifteen days to dispose of the government's petition for a rehearing. Acknowledging the significance of the case, the Court patiently answered all objections of the government's attorneys and concluded, "the decision could be no other than that which the learned judge of the circuit court below made, and which this court affirmed."[41]

McMains may have done little or no damage to the Maxwell title through his efforts in Washington, but he succeeded in taking much of the joy out of Springer's triumph. Discouraged with the course of events and frustrated that he could not free up time for scientific work, Springer told Charles Wachsmuth, "The worst thing is the inexcusable conduct of the U.S. officials in the Maxwell Land Grant case. After we have beaten them completely, on every possible point, a lot of cranks still persist in thinking they can do something to keep me in hot water, and think they can litigate the whole thing over again by another suit in New Mexico."[42]

The New Mexico case consumed two more years, culminating in a decision of the territorial Supreme Court in January 1889. As Springer had anticipated, the determination of the territorial courts was a relatively simple one; the case was deemed *Res Adjudicata,* already decided under jurisdiction of the U.S. Circuit Court of Colorado and the U.S. Supreme Court.[43]

Aftermath

Following the decision of the U.S. Supreme Court in April, 1887, an estimated five hundred families who had settled on the Maxwell claim faced the very real possibility of eviction. Within days of the 1887 decisions of the Supreme Court, the *Raton Daily Range* was warning of a "threat of war" over the Maxwell controversy.[44]

In a front-page headline on June 28, the *New York Times* proclaimed, "Settlers Ready for War: Trouble Brewing on the Maxwell Land Grant." Though

O. P. McMains (on horseback) and settlers on the Maxwell Land Grant, east of Raton, New Mexico, c. 1888. Courtesy of New Mexico State University, Archives and Special Collections, Audrey Alpers Photographs.

unattributed, both the article and the inflammatory circular it quoted bore the fingerprints of O. P. McMains. According to the article, guns had been purchased for the purpose of arming the settlers. A handbill circulating on the grant told settlers, "Come to Raton in force by the first day of next August, and if we cannot in mass meeting assembled, without a rumpus, get the attention of the Nation to the crime committed by Williamson in arbitrarily surveying public land in the Maxwell Grant, then let us by all means have a rumpus."

The armed rebellion did not materialize. In September President Cleveland attempted to calm the waters, furnishing Maxwell Company officials with a letter for publication. The president referred to the recent decision and declared, "The judgment of the Supreme Court of the United States on the subjects involved is authoritative and conclusive. Its judgment must be respected and obeyed." The president gave further assurance that violence would not be tolerated, stating, "Any unlawful overt act committed in pursuance of such counsel will, of course, be visited with the penalty appropriate to the crime."[45]

Despite such pronouncements, there were further confrontations among settlers, agents of the Maxwell Company, and officers of the law. The company pursued ejectment proceedings with some success, but there remained pockets of resistance. The hostilities reached their climax in August, 1888, when a confrontation between antigrant settlers and a handful of sheriff's deputies at Stonewall, Colorado, culminated in a gun battle and the deliberate burning of a barn and a hotel; two men died in the fracas.[46]

McMains did jail time for his part in the affair, but in the fall of 1888, he issued a broadside titled, "DEFIANCE and Contempt for that which is Contemptible." McMains stood firm on the Cox decision and proclaimed, "SETTLERS—Your case is won."[47] In fact, the case was anything but won. The futility of the cause was becoming increasingly clear, and most settlers had concluded that they would have to reach an accommodation with the company or leave the country. McMains continued to carry the fight against the Maxwell Company until his death in 1899, a shrill but ineffectual agitator.

In recognition of Frank Springer's work in the patent suit, the cash-poor company rewarded him with lands totaling sixteen thousand acres, "together with the rights of way and use of water for the necessary ditches above said land, to irrigate the same from the Cimarron and other streams tributary thereto."[48]

Springer doubtless believed in the cause for which he fought, and his interpretation of the matter found resounding confirmation in courts of law. There were, however, two sides to the controversy, each with valid points, and there were and are well-informed observers whose views more nearly coincide with those of McMains and George Julian than of Springer, Judge David Brewer, and the U.S. Supreme Court. On examination of the circumstances, it would appear that the case might have gone either way, but there were factors other than Springer's legal prowess to suggest the eventual result. Brewer, the judge in the Colorado circuit court, was a consistent defender of property rights, frequently siding with large commercial interests. In addition, the case was decided in an era beginning around 1865 and lasting until the Great Depression, in which many decisions of the Supreme Court were friendly to big business.

While Springer made some enemies among the antigrant forces in the course of the long struggle, he emerged a man of greatly increased reputation in the territory of New Mexico, in his profession, and among his associates in the Maxwell Land Grant Company.

CHAPTER 6

One Man's Destiny

The successful conclusion of the patent suit marked the beginning of the busiest phase of Frank Springer's career as an attorney and businessman. Springer continued to represent the Maxwell Land Grant Company, prosecuting numerous suits for ejectment and seeing his client through several controversies that had major implications for the financial health of the company. He also continued as solicitor for the Santa Fe Railroad and took on other clients who sought his services.

Even as Springer concentrated his energies on the practice of law, he remained true to his larger purpose in migrating to New Mexico, which was to accumulate wealth and achieve financial independence. Springer had been an investor from his earliest days in the territory; as his means allowed, he continued investing in promising enterprises.

Among the obligations motivating Springer's effort was the responsibility of providing for a growing family. Following the births of their four daughters, Frank and Josie Springer also had three sons, each of whom was named for a person admired by Springer. Edward Thomas Springer, named for the companion with whom Francis Springer had migrated from Maine to Iowa, was born on May 15, 1885. Henry Strong Springer was born on August 21, 1887, and Lew Wallace Springer, on October 12, 1890.

Springer had long wished to build a comfortable family home. Soon after locating in Las Vegas he had purchased property in the Hill Site addition in East Las Vegas—a neighborhood of stately multistory homes that in no way resembled the adobe dwellings around the old west side plaza. These elegant homes were occupied mainly by the families of merchants, railroad officials, and professional men. Springer had purchased the land in 1884 but did not build on the property until 1891; the structure was finished in 1893.

Children of Frank and Josephine Springer. Girls (clockwise from left): Ada, Laura, Helen, Eva. Boys (from top): Edward, Wallace, Henry. Courtesy of CS Cattle Company.

The Queen Anne–style home built at the corner of Washington Avenue and Sixth Street had rooms on three floors and was crowned with steep gables.

Perhaps in connection with construction of the new home, Josephine and the Springer children relocated to Los Angeles in 1892 while Springer maintained his headquarters in Las Vegas. Springer visited his family from time to time over the next few years, traveling with Josie and children to see the sights at San Francisco, Del Monte, Santa Monica, Long Beach, San Pedro, Santa Catalina, and Redondo, before returning to New Mexico and the certain demands of his business affairs and legal practice.[1] There is no evidence that this separation was a result of marital disharmony or that Springer contemplated a permanent move to California. The move may have represented an attempt on Josie's part to situate her family more comfortably, realizing that they could not expect to share much of her husband's time in either location.

The Maxwell Cattle Company

While Springer was personally interested in a variety of enterprises, the bulk of his resources were invested in cattle. In addition to building a herd on their own ranch, Frank and Charles Springer also were involved in several corporate stock raising enterprises in the 1880s.

The most significant of these was the Maxwell Cattle Company, organized in 1881. This venture was the cornerstone of Frank Remington Sherwin's strategy for exploitation of the grant. Despite reservations on the part of some of the company's Dutch bondholders, Sherwin organized a large-scale cattle operation, which was separately incorporated and to which an extensive lease of company lands was granted.[2] Sherwin was the cattle company's first president. Springer was elected vice president; Harry Whigham was secretary; and Henry Porter was designated as treasurer. Manly Chase was employed as manager of livestock operations.

The company was capitalized at $1 million, and 10,000 shares of stock were issued. The Dutch bondholders were under the impression that the Maxwell Land Grant Company—whose resources comprised the company's principal capital asset—would be the owner of all but a few shares of the stock. They understood that Chase and Porter would be assigned a few shares each as an incentive to efficient management of the cattle company's affairs.[3] The Dutch were puzzled to receive only 7,250 shares of stock in the Maxwell Cattle Company, and those inexplicably unsigned. The bondholders may have been further chagrined to learn that the remaining shares had been divvied up among the company's American directors. Ownership of 1,000 shares each had been conveyed to Sherwin and Porter, with Chase, Harry Whigham, and Frank Springer each holding 250 shares. These shares apparently were assigned to the named individuals for a consideration amounting to 30 percent of actual value.[4] This disposition of the cattle company's shares fed a growing distrust of Sherwin. The company faced an additional obstacle in raising capital with which to purchase cattle. With the aid of loans from Henry Porter and others, some significant purchases were made, but funding problems retarded the company's progress toward profitable operation.

In an attempt to find an administrator who could be trusted to do their bidding, the Dutch in 1883 obtained the services of Hans Mattson, a Swede who had immigrated to the United States. En route from a tour of duty as U.S. consul general for India, Mattson met with members of the bondholders' committee in the Netherlands and agreed to help remedy the company's affairs in New Mexico. He accepted an appointment as general manager of the Maxwell Company and traveled to New Mexico with members of the Dutch committee.[5] There the group met with Sherwin and negotiated his departure as president of the Maxwell Cattle Company and of the Maxwell Land Grant Company.

Sherwin's resignation came at a steep price, as investors agreed to buy out his interest at a cost of well over half a million dollars.[6] This was a staggering obligation for a company with a paucity of cash and severely limited options for raising more. To secure the obligation to Sherwin, the company placed some 5,800 shares of Maxwell Cattle Company stock in trust under Porter and Springer. Henry Porter succeeded Sherwin as president of the cattle company. In a move intended to improve the company's ailing financial condition, the directors increased the capital stock to $1.75 million.

Thoroughly aroused, the Dutch sent one of their own, Martinus Petrus Pels, to investigate and provide an independent report on affairs of the Maxwell Company. By midyear 1885 the Dutch investors had taken matters into their own hands, forming a strong committee to speak for the bondholders and giving Pels power of attorney. The committee remained optimistic, advising bondholders that the property would yield good returns once free of its present liabilities. The company's troubles were attributed to "the too great power originally given to Mr. Sherwin." As a remedy, it was suggested that the existing company be terminated and replaced by a new entity.

Mattson may have been of service in helping untangle the company's affairs following the Sherwin debacle, but by the end of 1886 he had returned to his home in Minnesota, and it was left to Springer, Porter, and others to effect a reorganization under which the company could remain a viable enterprise.

Mired in receivership and beset with troubles related to Sherwin's administration, the Dutch committee resolved to pursue a reorganization of the Maxwell Land Grant Company, but they had first to dispose of obligations to creditors and terminate the Maxwell Cattle Company. The cattle company was put out of its misery when satisfactory settlements were reached with Sherwin and associates and with the cattle company stakeholders. The cattle operation was reorganized as a department of the Maxwell Land Grant Company.

Sherwin's administration of the Maxwell Company was looked upon by the Dutch committee as a colossal mistake, setting back efforts to develop the property's resources and leaving the company with heavy encumbrances. One of Sherwin's interests did, however, bring an asset of lasting value to the area. Sherwin had an affinity for horses, and he believed that the climate of northern New Mexico was conducive to the raising of thoroughbreds.[7] As a personal business venture, he authorized Hume Webster, a respected English horseman, to purchase some fine breeding stock for shipment to New Mexico. Webster assembled twenty mares and a French-bred stallion. The horses were shipped to New York, moved to Springer by rail, and driven to Cimarron, where adobe stables had been specially built near the Maxwell headquarters.

Sherwin's vision of a successful breeding operation evaporated in the chaos of his financial difficulties, but some of his horses remained to improve breeding lines on area ranches for years to come. Following Sherwin's departure as president of the Maxwell Land Grant Company, the horses were sold at auction in 1885. A Kansas breeder bought the original stallion and several of the mares, but Frank Springer was also present, and he acquired two well-bred stallions and several fillies. Starting with this remnant from Sherwin's stable, Frank and Charles Springer established a successful horse breeding operation on their ranch near Cimarron.

Reorganization

The negotiations by which Sherwin was relieved of his role in management of Maxwell Company affairs culminated in a reorganization of the Maxwell Land Grant Company. According to the recollection of Henry Porter, "We sent Springer, who by this time understood the condition of affairs, over to Holland to see what could be done to straighten out matters and keep the company alive."[8]

It was Springer's responsibility to work out legal provisions for reorganization of the Maxwell Company. With a general plan in mind, he took passage for the Netherlands in the fall of 1887 and laid his ideas before the bondholders. The trip was not all business, as he also stopped in Bonn, Paris, and London to examine fossil collections, acquire specimens, and consult with scientific colleagues.[9]

Ralph Emerson Twitchell later interviewed Springer concerning his efforts in Amsterdam and explained the matter as follows:

> The existing company was a Dutch corporation which for practical reasons none of the parties wished to continue; and the Act of Congress then in force known as the Alien Law—which prohibited corporations more than twenty per cent of whose stock was held by aliens from owning land in the territories—precluded the possibility of an American corporation. Some other way had to be found for administering the property, and that quickly, to avoid failure of the negotiations. Without access to law libraries, or opportunity to consult with American counsel, Mr. Springer at Amsterdam devised and perfected the plan for vesting the property in an administrative Board of Trustees, representing all the parties interested, having the title and powers of owners in fee simple—such Trustees not only having a prior lien upon the property for securing the large issue of bonds at the same time created, but also the actual control, management and administration of the entire property, with full powers of disposition and sale—thus in effect taking the place of the corporation which had owned the property.[10]

By mid-January the plan had been approved by investors. Three million dollars in new "prior lien bonds" were issued and a deed of trust was executed, placing effective control of the property in the hands of a group dominated by eastern capitalists and attorneys—R. V. Martinsen, Samuel Parrish, Charles Fairchild, and Nicholas Thouron. Also appointed as trustees were

Frank Springer, H. M. Porter, and the Dutch envoy, M. P. Pels. Martinsen was elected president of the new board. James L. de Fremery later joined the board, replacing Fairchild, and was named vice president. A corporate office was established in New York and Pels was appointed general manager. Satisfied that the new organization was on sound footing, the district court discharged its appointed receiver, Harry Whigham, who was then employed as superintendent for the Maxwell Company.[11]

The years immediately following the reorganization saw a renewal of efforts to capitalize on the Maxwell Land Grant Company's assets in land, timber, coal, and other minerals.[12] Beginning in 1888, two ditch systems were developed in an attempt to improve potential farmlands and make them more attractive to immigrants. Coal mining was a paying proposition at times, but sales of coal and timber suffered from a lack of rail access. The company contracted with Grenville Dodge for a road to run from Trinidad to the Canadian River, then upriver for a distance of sixteen miles. The town of Catskill was laid out as the hub of the company's timber operation. The discovery of cement rock gave rise to the development of a small industry in the production of cement and related products. Gold mining activity remained at low ebb, producing only a few thousand dollars a year from scattered placer claims. Several years would pass before the Maxwell Company could initiate new efforts to exploit the mineral resources of the Baldy region.

Springer was still involved in mopping up minor actions associated with the patent suit when yet another challenge to the company's title arose. The Interstate Land Company brought suit claiming ownership to an area of some 60 million acres in New Mexico and the surrounding region. In retrospect, the claim appears absurd on its face, but enough land and wealth were at stake to attract speculators.

The dubious asset of the Interstate Land Company was its interest in a grant made to Stephen Julian Wilson by the Mexican governor of Coahuila-Texas in 1826 and taken up in March 1832, by John Charles Beales and Manuel Royuela.[13] Before the year was out, Beales had purchased his partner's interest. The claim eventually passed to Beales's children and was sold to speculators who formed the Interstate Land Company.

A lawsuit intended to establish the Interstate Land Company's title to the Colorado portion of the Maxwell Land Grant was filed in the Federal Circuit Court for Colorado in January 1888. Springer represented the Maxwell Company, arguing the matter before Judge Brewer. The Interstate

Land Company's claim was found to be without merit, but its attorneys filed an appeal in the U.S. Supreme Court. The case was heard on March 10, 1891, and on April 6, the Court returned a decision in favor of Springer and the Maxwell Land Grant Company.[14] In consideration of his extraordinary effort in defending the company against the Beales claim, Springer was awarded special compensation in the amount of two thousand dollars.[15]

Springer Takes the Chair

In the popular novel *Grant of Kingdom,* Harvey Fergusson applied his knowledge of the storied Maxwell Land Grant to the creation of a metaphor bearing on the phenomena of social and economic change. Fergusson concentrated on the period surrounding Lucien Maxwell's reign as owner of the property and lasting until the coming of the railroad—a phase Fergusson characterized as "primary pioneering."

For Fergusson, the notion of *destiny* helped explain the process of transformation that could be observed in frontier societies. The conquest of a wilderness like the one north of Taos initially required courage, physical hardihood, and resourcefulness. This created opportunity for a man like Lucien Maxwell or the fictional Jean Ballard. As long as the frontiersman's attributes were of primary importance, he prospered. When native tribes had been subdued and order established, he was supplanted by leaders whose qualities more nearly matched the demands and opportunities of the present. Springer saw this clearly enough, telling Charles Wachsmuth in 1881, "The development of this country which I have expected and waited for, is now at hand. . . . It brings a new element, a new class of men, and the situation necessarily calls for a man's best ability, and all his energy, industry and pluck."[16]

Springer lived some of the history evoked in Fergusson's story, but he epitomizes the era that was to come, when prevailing conditions placed a premium on education, legal prowess, rhetorical skill, planning, and shrewd management. Having served a long apprenticeship in which he had become familiar with its legal and fiscal affairs, Springer was uniquely prepared to lead the Maxwell Company out of the morass created by repeating cycles of mismanagement and into a new era of viability and progress.

Destiny is not a scientific notion, and it is unlikely that Springer put much stock in it as a force in his own life. He might sooner have gone along with a rough scientific equivalent, the Darwinian idea of natural selection.

Whether as a result of predestination, superior genes, his own vision and competence, or the vagaries of time and chance, Springer in 1892 came to head the enterprise that he had now served for nearly twenty years.

In the new Maxwell Land Grant Company's fourth year of operation, concerns surfaced regarding its high operating costs. Wary of seeing its valuable assets frittered away yet again, the Dutch committee asked that economies be implemented to conserve the company's meager revenues. At the same time, friction developed between the president and vice president of the American trustees.

For reasons of health, R. V. Martinsen, president of the trustees, took leave of his duties early in 1891, leaving the vice president, James L. de Fremery, to execute the duties of president. By summer Martinsen had recovered sufficiently to travel to Amsterdam to meet with Dutch officials of the company. The Dutch representatives urged measures to economize in administration of the Maxwell property, and Martinsen responded by suggesting that the company headquarters be relocated from New York to New Mexico, and that the offices of president and general manager be combined. In fact, a resolution dated July 29, 1891, directing a relocation of company headquarters to New Mexico and specifying that members of the executive committee would reside in New Mexico, was adopted by the trustees.

Martinsen inquired of de Fremery concerning his interest in assuming the combined duties of president and general manager. De Fremery answered in the affirmative, specifying that he would accept a salary of $12,000 per year, or at least $10,000.[17] Finding objections among the other trustees to the idea of combining the offices, Martinsen inquired concerning de Fremery's willingness to accept the presidency alone, at a salary of $7,000, with Pels remaining as general manager. In either case, Martinsen would voluntarily yield the presidency. De Fremery declined to accept the presidency with Pels remaining as general manager and struck out in pursuit of his own agenda, meeting with the Dutch directors in Europe after Martinsen had returned to America.

Meanwhile, Martinsen changed his mind about resigning the presidency and cabled de Fremery in the Netherlands to this effect. Martinsen expected that this would end the matter of de Fremery's consideration for the office of president. Martinsen was highly chagrined when, on August 9, he received a cablegram from de Fremery, informing him that the Dutch directors had approved de Fremery's appointment as president.[18] To Martinsen, two things were amiss in this information: Martinsen had not resigned the presidency, and did not now intend to do so, as he had informed de Fremery; and while the American trustees were sensitive to the wishes of the Dutch directors, the selection of the trustees' president lay with the trustees themselves, and not with the Dutch. With the possible intent of distancing himself from Martinsen and his administration, de Fremery resigned the vice presidency, whereupon Frank Springer was elected to that office.

After receiving de Fremery's cablegram, Martinsen met with the "western trustees"—Springer, Porter, and Pels—and found them supportive of his position. One can also read into the reaction of these trustees a distinct aversion to the notion of de Fremery as head of the company. On September 10, 1891, Springer wrote to Samuel van Houten to inform the directors in Amsterdam of the strong support for Martinsen which, according to Springer, included not only the western trustees but also two of those residing in the East. Springer outlined plans by which savings would be realized, Martinsen relocating to Denver but requiring no office there, and Pels carrying on as general manager.[19]

While Martinsen was more than willing to continue as president of the American trustees, it became increasingly clear that he had lost the confidence of the Dutch. Battle lines were drawn, with Martinsen and the western trustees on one side, de Fremery, Parrish, and Thouron on the other. Martinsen had numbers on his side, but he had little support in Amsterdam, and it was becoming increasingly apparent that he would be unable to unify the trustees. On his own initiative, Martinsen quietly proposed to Pels and Springer that he resign the presidency in favor of Springer, who had the confidence of the Dutch and who, in conformity with the resolution of July 29, already lived in New Mexico.

Martinsen floated the notion to Pels in a telegram dated December 9, 1891. Surprised at this turn of events, Pels replied, "Do not wish you to resign. What is your reason?" At the same time, Pels expressed his approval of Springer as a possible candidate for the presidency. The exchange continued on December 10, Martinsen restating his intention to resign in favor of Springer and noting that the executive committee—composed exclusively of eastern trustees—was intent on forcing him from the board one way or another.

Frank Springer had not sought the presidency of the Maxwell Land Grant Company, but when the opportunity came his way, he seized the moment. On December 14 Springer wrote to M. P. Pels, mapping out a strategy by which the proposal to elect him president might be achieved, at the same time shifting the locus of power from the board's eastern members to a group dominated by westerners.

Finding the already adopted resolution of July 29, 1891, ideally suited to the purposes of the western trustees, Springer urged upon Pels the idea that they "take the resolution of July 29 as having been *unanimously adopted,* as they now insist, and treat it as a *fait accompli.*" With the resolution as a springboard, Springer laid out a course of action.

> Then by virtue of that resolution the "New York office shall *terminate* on the first day of January, 1892, and the *only office* of the Board of Trustees of the Maxwell Land Grant shall be located at Raton &c." Martinsen wants to resign December 31st, 1891, in which case the Vice President [Springer] becomes the acting President, and the next part of the resolution would be complied with, viz that the "President . . . shall reside within the Territory of New Mexico." We should have to elect a new resident Treasurer. The next thing would be the members of the Executive Committee, whom the resolution provides must all reside within the Territory. This would be a strong ground for an *immediate* change of at least one member of the Board of Trustees, so that we could have a new Trustee, resident in New Mexico. This could be John [Jan] van Houten.[20]

In short, it did not take Springer long to find a multitude of virtues in the July 29 resolution which, only a short while ago, the western trustees had seemed quite willing to abandon.

Jan van Houten had recently come to New Mexico to pursue business opportunities on the Maxwell grant. He was an energetic and well-regarded businessman, and he was the son of Samuel van Houten, a leading member of the Dutch Maxwell Land Grant committee. It may have crossed Springer's mind that Jan van Houten's nomination as a trustee would not hurt the chances of this plan when offered to the Dutch for their consideration. Springer remained in frequent communication with Pels and Martinsen, laying the groundwork for a change in management of the company along the lines suggested by Martinsen.

Springer's ideas were adopted, and events closely followed the scenario he had laid out. The Maxwell trustees met in New York on February 1–2, 1892, and Springer was elected president without serious opposition. The board further resolved that the company's New York office would terminate as of March 1, 1892, with the company headquarters relocated to Raton.[21]

To complete what amounted to a watershed change in administration of the company, all of the eastern trustees shortly submitted their resignations, and on March 21, 1892, three new trustees were appointed. Joining the board in addition to Jan van Houten were Francisco Manzanares and J. M. Cunningham, both of whom were residents of Las Vegas and associates of Springer. With the change in board membership, the locus of power had shifted entirely to New Mexico and Frank Springer.

Springer assumed the presidency determined to reverse the company's dismal financial record and justify the confidence of the Dutch directors. In a meeting stretching over six days between May 9 and May 20, 1893, he led the trustees through a comprehensive review of the company's operations. The trustees heard reports on the Springer and Vermejo ditch systems, neither of which had performed up to expectations. The board voted to dispose of the Vermejo system and, on Pels's advice, rejected a proposal to involve the company in plans to build a dam and reservoir on the Cimarron River. On May 11 Springer conducted a review of timber and cattle operations. The board eliminated the position of manager of the cattle department and placed the operation on a strict budget. On May 18–20 the Maxwell farm and ditch operations were further scrutinized. The board reviewed mining operations, land sales, recent and pending litigation, and relations with the fledgling Maxwell City Land and Improvement Company. Financial procedures were tightened and reporting requirements made more rigorous.

The meeting was like no other the board had seen, but Springer did not let up in 1894, holding meetings on eight days between March 28 and April 6. The board's attention was focused on plans to stimulate mineral production and resulted in the adoption of new regulations for mining and prospecting. The trustees continued in this manner for several years as Springer attempted to jump-start development of the grant's ample but largely untapped resources.

Springer believed that there was great promise of a return on the property's extensive deposits of coal. He pushed the development of coal and coke production and pursued the construction of railroads to transport products to market. In early 1893 Springer published a twenty-three page memorandum in which he argued for a substantial investment in the grant's coal resources.[22] The argument fell on deaf ears in Holland, but Springer did not give up the notion that coal mining would one day lead to a breakthrough in the company's heretofore anemic financial performance.

As president Springer retained responsibility for legal representation of the Maxwell Land Grant Company and added to it a much-increased interest in fiscal and operational concerns. This entailed more frequent communication with the general manager, Pels, and with the Dutch bondholders' committee. In 1895 a significant change in management occurred when Pels resigned and Jan van Houten became vice president and general manager. Pels had been an abrasive presence in relationships within the company and without, and his departure likely was greeted with relief by Springer and others. On the contrary, the relationship between Springer and van Houten was cordial and mutually respectful.

By 1895 Frank Springer was a busy man and a somewhat frustrated one. Having found New Mexico a place of opportunity, he had applied himself diligently, but after twenty-two years he could only feel that the cherished goal of financial independence had eluded him. Weary, out of sorts, and thoroughly discouraged, Springer revealed the depths of his despair to his friend and collaborator, Charles Wachsmuth: "I have not been well this winter—more from trouble of mind, however, than of body. My business cares and anxieties have weighed heavily on me, and at times I have been pretty well discouraged. I am almost forced to believe that the best 20 years of my life have been thrown away, so far as permanent results are concerned—and this—I am sorry to say, largely through others whom in excess of confidence and good nature, I have helped and trusted, and who have in every single instance failed me when the time of need came. A few years ago I could see my way ahead to a moderate independence which would allow me to follow my inclinations for the rest of my life. But now I cannot see it at all, and I shall probably have to give up my ambitions, and spend the rest of my life in the treadmill of making a living." Springer also complained of trouble with his eyes and lamented that he could not at once repair to the quiet refuge of Wachsmuth's home and museum. He wrote in conclusion, "I am making plans for my family to come back from California in the summer, and live here again. The lonely life I have been living here for the past three years has become intolerable. Since last September I have been able to spend only 10 days with them. Please excuse this not very pleasant letter. I will try to write more cheerfully another time."[23]

Springer was not generally given to such candid expressions of his own discouragement. His family did return to live in the new home on Sixth Street in East Las Vegas and this may have lifted his spirits and enabled him to recover a greater sense of perspective. In any event, Springer worked through the crisis as he had worked through holidays, infrequent vacations, and the births of his seven children. The demands on Springer's time and energy remained heavy throughout the ensuing decade, but outwardly at least, he remained a highly successful business and professional man, respected by peers and admired in his community, and in time, he had ample reasons to assume a more cheerful countenance.

CHAPTER 7

Wachsmuth and Springer

In reviewing Frank Springer's career from a vantage point much nearer its end than its beginning, a colleague remarked that he had been leading a double life: "(1) As a lawyer and man of affairs in New Mexico, which has been his residence and home, and (2) As an original investigator and author in the science of Paleontology, in which he has produced a number of elaborate works published by Harvard University and the Smithsonian Institution at Washington—works which, as stated in *Popular Science Monthly,* 'are better known in London than in New Mexico where he resides.'"[1] Science may well have been Springer's first love, but during the most active years of his career, it necessarily occupied a subordinate position. Paleontology provided a welcome diversion from activities that were more mundane but that paid the bills and provided funds for investment. Fossil study usually had to be squeezed into late nights and lags in legal and business activity, of which there were few.

During the last thirty years of his scientific career, Springer was a solo practitioner, but from the beginning of his efforts at serious research and for more than two decades, he carried on a scholarly collaboration with Charles Wachsmuth. The relationship was characterized at various times by mutual respect, affection, admiration, anger, frustration, guilt, and resentment. Despite some turbulence, the connection was one that never broke or seemed in serious danger of breaking. The collaboration produced an imposing shelf of materials of enduring value to scientists interested in the fossil crinoids.

In some ways they made an unlikely pair. Charles Wachsmuth was a German immigrant nineteen years older than Springer. Frail from his youth, Wachsmuth made a success at business in Burlington, Iowa, before retiring to a quiet life that included some travel and pursuit of an avocational interest in the science of paleontology. While Springer sought his fortune

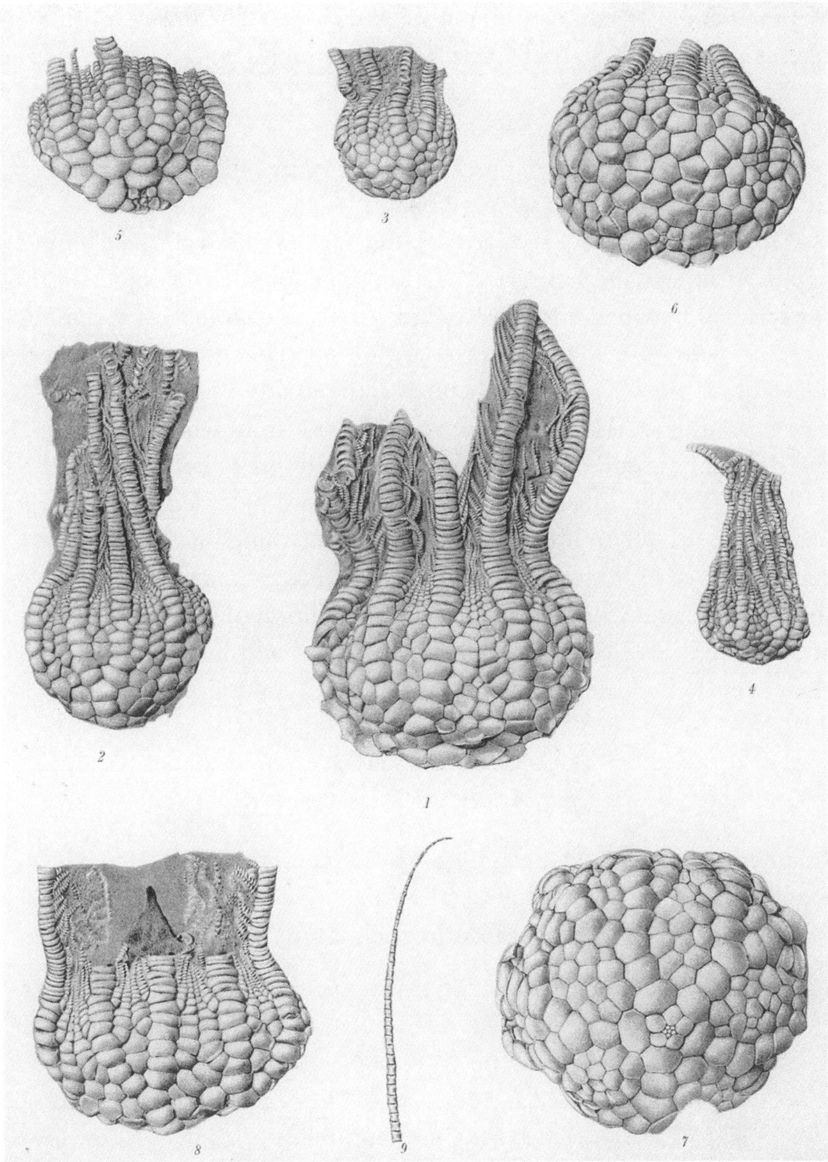

Uintacrinus socialis. *From Frank Springer,* Uintacrinus: Its Structure and Relations *(1901).*

in New Mexico, Charles Wachsmuth grew an unruly head of gray hair and whiskers, nursed his ailing heart through numerous episodes, and busied himself with books and fossils.

What Springer and Wachsmuth had in common was an intense, consuming interest in the fossil crinoids found along the limestone bluffs of

Burlington and in numerous localities worldwide. The object of their interest was a class of marine invertebrates commonly known as sea lilies or feather stars—a life form seeming to inhabit a region somewhere between flora and fauna but in fact belonging to the Echinodermata—a phylum of otherworldly sea creatures counting among its members the starfishes, sea urchins, sand dollars, and sea cucumbers.

Occurring in widely scattered deposits around the world, fossil crinoids appear in almost limitless variety. Part of their appeal as an object of study is explained by Charles Rollin Keyes, a scientific colleague of Wachsmuth and Springer: "For the study of evolution no class of organisms is better adapted. All the changes in structure can be clearly traced from one species to another, from genus to genus, and even from family to family. As the skeletal parts are composed of regular plates, definitely arranged, and frequently highly ornamented, the crinoids have recorded all the marked changes in the physical conditions of their surroundings, both as regards time and place."[2] Perhaps owing to the influence of Louis Agassiz, who rejected Darwin's teachings concerning the evolution of species, Wachsmuth and Springer gave little attention to this aspect and instead concentrated their attention on the morphology, or classification, of the Crinoidea.

An Accidental Scientist

In an essay titled "A Man's Leisure Time," conservationist Aldo Leopold extolled the virtue of hobbies and recalled one of the more interesting hobbyists of his hometown of Burlington, Iowa:

> When I was a boy, there was a German merchant who lived in a little cottage in our town. On Sundays he used to go out and knock chips off the limestone ledges along the Mississippi, and he had a great tonnage of these chips, all labeled and catalogued. The chips contained little fossil stems of some defunct water creatures called crinoids. The townspeople regarded this gentle old fellow as just a little bit abnormal, but harmless. One day the newspaper reported the arrival of certain titled strangers. It was whispered that these visitors were great scientists. Some of them were from foreign lands, and some among the world's leading paleontologists. They came to visit the harmless old man and to hear his pronouncements on crinoids, and they accepted these pronouncements as law. When the old German died, the town awoke to the fact that he

> was a world authority on his subject, a creator of knowledge, a maker of scientific history. He was a great man—a man beside whom the local captains of industry were mere bushwhackers. His collection went to a national museum, and his name is known in all nations of the earth.[3]

The subject of Leopold's reminiscence was Charles Wachsmuth.[4] A native of Hanover, Germany, Wachsmuth was plagued with poor health as a child, and was thus unable to obtain the requisite education to follow his father into the practice of law.[5]

Wachsmuth instead took up a career in business, and in 1852 was sent to New York as the agent of a Hamburg firm. The climate proved unfavorable, so with the encouragement of friends, Wachsmuth traveled to Iowa and settled in Burlington, where he established a grocery business. On June 3, 1855, he married Bernardina Lorenz, a native of Wachsmuth's hometown.[6] The business did well, but Wachsmuth's health problems continued and he was advised to commence a program of outdoor exercise. This he did, and in the course of long walks around Burlington, he began picking up interesting specimens among the fossils that were plentiful along the bluffs above the Mississippi River.[7]

Wachsmuth soon amassed an impressive array of crinoid specimens. By the time Louis Agassiz of Harvard visited Iowa on a lecture tour in 1864, the collection was sufficiently well known that the professor stopped in Burlington to see it and visit with Wachsmuth. This brief encounter marked the beginning of an extended relationship. In 1865 the Wachsmuths disposed of their business in Burlington and traveled to Europe, also making a visit to Agassiz at his Museum of Comparative Zoology at Cambridge.

When Frank Springer located in Burlington following his graduation from the state university, it was almost inevitable that his interests would lead him to the crinoids, and to Wachsmuth. By the early months of 1869, the two men were at least acquainted.[8] By 1871, they were hunting fossils together.[9]

Soon after Springer's departure for the Southwest in January, 1873, Louis Agassiz passed through Burlington on his way home from a trip to the West Coast and visited Wachsmuth and his collection a second time. As related by Charles Keyes, "[Agassiz] was greatly surprised at the enormous growth of the collection since he had last seen it, and, struck by the beauty and perfection of the specimens, he intimated that he

Louis Agassiz at blackboard. Courtesy of Schlesinger Library, Radcliffe Institute, Harvard University, Neg. A70-72-8.

was anxious to procure the collection for Cambridge, at the same time expressing a desire to have Mr. Wachsmuth go with it and take charge of all of the crinoids in the museum."[10] Negotiations ensued and the collection was purchased for the then-considerable sum of six thousand dollars.

Wachsmuth accompanied the collection to Cambridge, where he was employed by Agassiz as an assistant, with the responsibility of unpacking the specimens and placing them on exhibit. During the time with Agassiz, Wachsmuth was encouraged to go beyond his work as a collector and casual student of the crinoids and publish the results of his researches, rather than see his work redound to the credit of others.

Before the year was out, Agassiz had died and Wachsmuth had left Cambridge. In spring of 1874, the Wachsmuths embarked on travels in Europe, Asia, and Africa. Upon their return to America, Wachsmuth began collecting fossils again, and over a period of several years he assembled a collection that reportedly was even larger and more desirable than the one deposited in Agassiz's museum at Harvard. Wachsmuth eventually turned to serious investigation of the crinoids and got down to the tedious business of reducing his findings to writing. He sent drafts of his first tentative efforts for Springer to critique. In time the two men began collaborating on brief scientific papers.

Joint Authors

The first two collaborative publications undertaken by Wachsmuth and Springer were short papers—an eleven-page article published in the *American Journal of Science* on the genus *Belemnocrinus,* and a forty-two page paper, "Transition Forms in Crinoids," published in the *Proceedings of the Academy of Natural Sciences*. In March 1877, Springer made the first overture leading to a more extensive collaboration, involving a revision of the classification scheme for the crinoids. "If it should happen that by next winter I am in a position to work with you," he wrote, "I propose that we undertake the classification of the Palaeocrinoidea. We may not be able to do it *perfectly,* but we can make a great advance on anything yet done and add materially to the comprehension of the subject. This would involve a thorough revision of many genera, and probably the creation of new genera to receive certain forms of *Platycrinus* and *Scaphiocrinus*. With your knowledge and our joint practical experience, I feel sure we could do a good work."[11] By mid-December Springer had yet to make any progress on the work. He did, however, return to Burlington to work with Wachsmuth for several weeks in 1878, consulting essential reference materials and availing himself of the all-important collection of fossil specimens.

Wachsmuth apparently did the bulk of the drafting, and Springer edited. Springer worked to improve the form and style of the text, but such matters were of little interest to Wachsmuth. Springer conceded the primacy of careful analysis and sound scientific conclusions, while insisting that equal attention be given to the manner of expression. "I am not in favor of verbiage," Springer wrote, "but we must not trim too closely. I want to use words enough to make our article not only *correct* as a matter of fact, but also logical, clear, and attractive in composition, so far as I am able to do so."[12]

The work was begun with enthusiasm, but a persistent and perhaps inevitable obstacle was soon evident, as Springer found himself caught between his commitment to the work with Wachsmuth and the demands of his law practice and business involvements. In a gesture that came to typify the working relationship, Springer offered financial support in lieu of the labor he could not provide. "As you will have so much extra work in this in which I *could* help very much if I were there, I propose that if you get up the figures &c. I will pay for the whole of the illustrating of this paper," he told Wachsmuth. "I think it would be fair for me to do this."[13]

Part 1 of the *Revision of the Palaeocrinoidea* came out in 1879, but Springer's difficulties in getting to the work continued. In November, 1880, he wrote a long and rather pained letter, telling Wachsmuth, "I have felt for a long time that I was not doing my share of the work, and that it is probably unjust to you that I should have any share of the credit. I feared it would be so from the start. My interest in the subject was so great that I was led, unwisely perhaps, to undertake to cooperate with you in it. I still feel the same, or even greater interest, and I am anxious to do all I can, now that I have got into it."[14] Yet he could not see past a multitude of conflicting obligations, and he asked Wachsmuth's continued forbearance.

Springer blamed "the fates" for his inability to attend to the fossil work, but fate may have had less to do with Springer's harried condition than did his own consuming drive to attain the success for which he had forsaken a comfortable life in Burlington. Springer continued to express optimism that the pressure of his business commitments would recede and allow him to give more of his attention to science, but by the end of 1882 he was again pleading overwork and soliciting Wachsmuth's indulgence.[15]

In spite of the inherent difficulties, the collaboration was an immensely productive one. Between 1877 and 1886, Wachsmuth and Springer produced more than eight hundred pages of text, including the four installments comprising the *Revision of the Palaeocrinoidea,* published in the *Proceedings of the Academy of Natural Sciences of Philadelphia.* These and other publications that followed over the next decade may have been unknown to Springer's associates in Las Vegas and Wachsmuth's nearest neighbors in Burlington, but they were read with interest in Cambridge, Washington, London, and Bonn, where learned scientists could appreciate their worth.

The Crinoidea Camerata

With the publication of the *Revision of the Palaeocrinoidea* and related works, Wachsmuth and Springer attained the standing of known professionals in the field of study to which they had devoted themselves. They established alliances, exchanged scientific material with colleagues, and sometimes entered into the petty rivalries for which academia is well known.

Springer's continuing interest in the work was evident when, in 1886, he purchased an acre of land at Corey's Bluff, a site rich with fossil deposits near Crawfordsville, Indiana.[16] His commitment to the collaborative relationship was further evident when he wrote Wachsmuth to say, "I am going to enclose with this or perhaps next mail, a rough sketch of ground plan and general specifications for a fire proof building in your yard, and you will greatly oblige me by giving it enough attention to be able to give me pretty full information as to the cost by the time I pass through there, early in February. We *must* build it, and I propose to make a fair contribution to the cost of it."[17] The small brick museum was built beside the Wachsmuths' home, Springer doubtless providing a fair share of the cost, if not all of it.

In the early years of the partnership Springer made regular pilgrimages to Burlington to work with Wachsmuth, but as his responsibilities in New Mexico increased, these working visits became shorter and less frequent. Springer's work for the Maxwell Land Grant Company, the Santa Fe Railroad, and other clients involved extensive travel in the United States and abroad. He rarely left home without an agenda that included pursuit of scientific as well as legal matters.

The relationship between Charles Wachsmuth and Springer may not always have been the most congenial, but the effort was more than justified by the accolades of scientific colleagues on both sides of the Atlantic. Having survived the *Revision of the Palaeocrinoidea,* and having derived considerable satisfaction from its favorable reception, Wachsmuth and Springer found encouragement to embark on an exhaustive monograph on the Paleozoic crinoids as arranged in three primary groups: the Camerata, the Inadunata, and the Articulata.

Both men enjoyed the friendship of Alexander Agassiz, who had succeeded to the directorship of the museum founded by his father. In commencing the new investigation, Wachsmuth and Springer also initiated a purposeful cultivation of the younger Agassiz. Wachsmuth wrote to Agassiz in July 1887, and Springer followed up in early 1888, citing his intellectual debt to Louis Agassiz and telling about the project at hand. "Our work is the result of an undertaking in which Wachsmuth furnishes the brains and I the money, and both, I think, a fair equivalent of enthusiasm," he wrote. "We are now fairly started on a monograph of the Palaeozoic Crinoids of North America, and have an excellent artist at work on the drawings under Wachsmuth's own eye."[18] Springer expressed interest in acquainting Agassiz with the work and gave notice of forthcoming requests for loan of specimens.

Wachsmuth and Springer continued to massage the relationship with Alexander Agassiz, and in early 1890, the result that they must have hoped for came, for Agassiz offered to publish the new monograph in the *Memoirs of the Museum of Comparative Zoology.* At this point, the work was some four and a half years from completion and more than seven years from actual publication. Agassiz may not have realized that he was undertaking a role that would place him at the center of controversy between two proud and willful collaborators.

As the work grew tedious, old tensions surfaced. Springer became an unreliable partner as business commitments more frequently interfered with planned trips to Burlington. In place of his active presence, the proffered apologies wore thin. In lieu of his participation, Springer supported the project financially and promised to attend to his share of the writing at the earliest opportunity. Wachsmuth took no pains to conceal his consternation, and Springer responded with expressions of remorse, citing the overwhelming character of his professional obligations and noting that he had been cautioned by his doctor concerning his work habits.[19]

The work originally contemplated was a project of staggering proportions. By the end of 1893, it had become apparent that the original concept was too large and limits would have to be imposed in order to accommodate Springer's obligations and Wachsmuth's failing health. "The work, as you perhaps know," Springer advised Agassiz, "only includes the Camerata crinoids, but these are by far the most important, both in themselves as bringing to light new facts and relations, and with regard to the value of the special knowledge of Wachsmuth."[20]

Wachsmuth's condition was fragile, but he could work at intervals. By the fall of 1894 he was able to submit the finished manuscript to Agassiz with a note of explanation concerning the delay in its transmittal. "Our monograph on the Camerata of North America is at last completed," he wrote. "I have forwarded to you today by Adams Express (charges paid) 3 packages of plates, and one package of manuscript, which I hope will come safely to hand. That the MS. was not sent sooner was not my fault; my part of the work had been finished last spring, but Mr. Springer was too much engaged to attend to the reading of it, and finally our type writer got sick, and caused another delay of six weeks."[21] It was at this point that Agassiz's work as referee began in earnest.

Having received the first proof sheets from the press at Harvard, Springer found it necessary to address Wachsmuth concerning the style of the text—a continuing source of tension between the two. Springer was particularly chagrined to read Wachsmuth's clumsy edit of Springer's introduction to the monograph and expressed his dismay. "As it stands," he concluded, "it contains a number of errors in style, composition and grammar, for which I should not like to be held responsible, and I am sure *you* would not if you knew the comment that any good judge of English composition would make on reading it."[22] Springer asked to see proof copy before it was approved and returned to the press for printing—a request that doubtless rankled Wachsmuth. Springer also complained of the matter to Agassiz, seeking his help in dealing with Wachsmuth's increasing dogmatism and irritability—behaviors Springer attributed to a variety of infirmities.[23]

Perhaps grudgingly, Wachsmuth complied with Springer's request, sending the proofs on for his inspection before returning them to the press. Springer undertook to improve on the manuscripts sent him from Iowa, telling Wachsmuth: "I am going to take the liberty of making a few corrections in the English, for there are some expressions that I would not like to have stand with my name attached; and I am sure—although I know you have a prejudice against my style of composition—that you also wish the text to be in fairly acceptable English. As for changes making you 'ridiculous in the eyes of the printer,' I will say that I have not spent $10,000, nor you twenty years of time, in bringing out this work, for the purpose of pleasing the printer." Springer went on to lecture his collaborator on distinctions among the words, *obviously, evidently, probably,* and *apparently.* "As to *pari passu,*" he wrote, "my objection to it is that it is pedantic, and unnecessary. It is not at the present day considered good style

Charles Wachsmuth in his study at Burlington, Iowa. Courtesy of Department of Paleobiology, National Museum of Natural History, Frank Springer Collection.

to use Latin expressions where there are good English equivalents, as there is in this case. It is a matter of taste purely, as to which I seem lately to have become a poor judge."[24]

After two months of acrimonious correspondence, the two men had exhibited their mutual antagonism and Springer wrote to convey sentiments that were only partly conciliatory. The relationship remained intact, but the collaborators continued to work at cross purposes, Springer occasionally enlisting Agassiz's help in conveying criticisms to Wachsmuth, Wachsmuth resisting Springer's insistence on stylistic perfection.

Despite the difficulties, the massive *North American Fossil Crinoidea Camerata* moved toward publication. As Springer had intimated, no expense had been spared, either in preparation of the work or in its production under auspices of the Museum of Comparative Zoology. With its 800 pages of text, 83 plates, and 1,500 illustrations, the manuscript was many months in the process of typesetting, proofing, and printing. Although the work had been in press well over a year, bound volumes had not been delivered by the end of 1895.

The End of an Epoch

Wachsmuth and Springer continued to attend to the details of publication and dared to consider further efforts in pursuit of the ambitious project they had first envisioned. Wachsmuth, however, grew weaker, to the point that he could work only sparingly. Springer, as usual, was immersed in business and belatedly learned of the gravity of Wachsmuth's condition. In January 1896, he expressed himself with the concern of an old and dear friend, and in a sadly familiar manner. "I have just returned home from another RR business trip, and am very much grieved to learn that you have been so ill," he wrote. "I have been trying all winter to plan so as to go and spend a few weeks with you, but it seems to be impossible to arrange it."[25] Springer cited complications arising from a reorganization of the Santa Fe line, a controversy concerning its connecting link to California, and ongoing demands of the Maxwell Land Grant Company, concluding, "It is not for want of the desire to do so that I do not go to see you—but the fates seem to be against me at the very time I wish to go, because I know you need me."

Wachsmuth soon had no further need of aid and comfort. Following Wachsmuth's death on February 7, Springer made time to go to Burlington.

There he performed a service requested of him some years earlier, addressing the throng gathered to honor Wachsmuth's memory. The funeral was held at the Wachsmuth home where, the *Burlington Daily Gazette* reported, the crowd of mourners far exceeded the capacity of the house.

Springer told the assembled mourners of Wachsmuth's long struggle with pain and disease, and of the near fatal episodes suffered by him. He praised Wachsmuth as one who set an example of "patient suffering and cheerful fortitude, of high courage, and of iron will which rose superior to his physical infirmities." In Springer's estimation, Wachsmuth's character had much to do with his approach to science: "A man of warm and generous nature; of unyielding fidelity to his convictions of right and truth; tenacious of his opinions, bold in avowing and sturdy in defending them, yet always ready to surrender a theory and accept the proven fact; scrupulously exact in regard to all his obligations toward his fellow men; his life among you has been an honor to this community, and to the state to which it belongs. It has been a quiet life, without ostentation or display of any kind, even in things of which he had a right to be proud."[26] Wachsmuth was further memorialized in the *Bulletin of the Geological Society of America,* and in other scientific journals. In a notice of the forthcoming publication of *The North American Crinoidea Camerata,* Charles R. Keyes reviewed the achievements of Wachsmuth and Springer and characterized their long and fruitful collaboration as "an epoch in the history of American science."[27]

When the *Crinoidea Camerata* appeared in May 1897, it was an impressive contribution—two large quarto volumes of text and an atlas containing the many plates and drawings, all beautifully printed and bound. In a note preceding the text, Springer paid a final tribute to his collaborator, saying, "It is a source of extreme regret that my learned colleague and long-time friend, Dr. Charles Wachsmuth, did not live to see the publication of this Monograph, to which he had devoted so many years of assiduous labor. Never a robust or healthy man, his last few years were almost a continual struggle against disease."[28]

To Frank Springer, Wachsmuth's death meant the loss of a teacher, friend, and confidante, as well as a partner in highly productive and rewarding scientific ventures. Despite their difficulties, the two men valued one another immensely, and it is doubtful that each working alone could have accomplished anything near what they achieved together.

Springer maintained the library and collection at Burlington for some fifteen years following Wachsmuth's death, enjoying Bernardina

Wachsmuth's hospitality and assisting her financially. He also continued in the line of research begun with Wachsmuth, preparing new papers for publication as his energy and resources allowed. The loss of Wachsmuth and a heavy involvement in business and legal affairs necessarily brought about an immediate decline in Springer's productivity as a paleontologist. There was, however, no loss of interest in the work, and there would be other substantial achievements before he was done.

CHAPTER 8

Springer at the Bar

During the three score years following his graduation from the State University of Iowa, Frank Springer accomplished much in law, business, and science. He also became a major contributor to the development of education and the arts in his adopted territory and state. Only twice did he hold elective office and he did not seek political appointments, but he was an interested and involved member of the Republican Party who freely expressed himself on legislative matters and sometimes worked behind the scenes to influence decisions. He was a family man to the extent of choosing as his wife a woman who must have given careful attention to the raising of their seven children during his frequent absences, and he was supportive of his children in their pursuits as young adults.

Springer attained preeminence in an astonishing array of activities, but the bedrock of his success was his ability and diligence as a practicing attorney for more than thirty-five years. In the decade following his victory in the Maxwell patent suit, Frank Springer appeared before the U.S. Supreme Court several times, achieving widely publicized victories in issues decided at that level. Springer's practice regularly involved litigation in the district courts and the Supreme Court of the territory of New Mexico; here too he achieved a remarkable record of success. Following his brief experience as a prosecutor in Burlington, Springer's practice was chiefly in the realm of civil law.

In addition to his work on behalf of the Maxwell Land Grant Company, Springer continued to represent the Atchison, Topeka and Santa Fe Railroad Company, serving as assistant attorney under Henry Waldo. By his own reckoning, Springer handled some 106 railroad cases between 1879 and

1898, with the result that "only a single one was adversely decided, and that a minor one contested against the advice of the law department."[1]

Springer had a brilliant mind, extraordinary rhetorical skill, and a strong work ethic. Asked to disclose the keys to his success, Springer pointed to two guiding principles:

1. Thorough study and perfect preparation for the trial. Leave nothing to chance that foresight can provide against. A program for the case is put in systematic notes and assembled in logical order and with mathematical precision.
2. The ability, acquired by experience, to discriminate between the essential and the non-essential. Don't waste energy on immaterial points, unless for some strategic reason. In almost every lawsuit there is some one crucial point or proposition of fact or law upon which the decision must ultimately depend.[2]

According to the New Mexico attorney and historian Ralph Emerson Twitchell, Springer had "a deep-seated pride in the Profession of Law as a great intellectual arena, and while in active practice rejoiced in success for its own sake, and had a corresponding horror of defeat."[3] Springer's moral refuge was in the law, and he was a strong believer in the adversary system of justice. Of the essence of legal practice, he told his colleagues, "The lawyer is first and last, and above all, a representative man. Notoriously incompetent to manage his own affairs, every fibre of his being is aroused to zeal in the cause of those whose interests have been confided to him."[4] In Springer's case, the matters confided to him included questions involving large enterprises and high stakes, so that the pursuit of his clients' interests was highly remunerative as well as personally satisfying.

The Court of Private Land Claims

One result of Springer's success in the Maxwell patent suit was that he, of necessity, became an authority on land grants and the laws relating to them. His services were suddenly in demand, as other land grant claimants facing legal tests wanted the man who had made the highly publicized Maxwell claim stick. In addition, Springer was in a position to speak with authority on the subject of land grants and public policy concerning their disposition.

In 1889, possibly as a result of increased stature following his success in the patent suit, Springer was elected president of the New Mexico Bar Association. Although the Maxwell patent case had been resolved to his satisfaction, Springer remained very much interested in the issue surrounding the many unresolved claims. On the occasion of his valedictory speech as president, he chose to address the bar on the subject of "land titles in New Mexico." Speaking to his colleagues on January 7, 1890, Springer reviewed the history of land grants in the Southwest and spoke of the chaos and economic loss resulting from the government's failure to resolve unsettled land titles. Having held his fire through active phases of the Maxwell litigation, he took the opportunity to blast Surveyor General Julian for what he called "a raid on New Mexico land titles" and to offer a highly uncomplimentary review of Julian's job performance.[5]

Springer doubtless derived satisfaction from his declarations on the transgressions of the surveyor general, but the greater significance of his address was in calling attention to the need for a more effective means for settling disputed titles. Springer's remarks were widely disseminated in the territorial press; upon action of the Bar Association, the speech was also printed for distribution to both houses of Congress.

An earlier article by Julian and an opposing one by Stephen W. Dorsey in the *North American Review* had contributed to public awareness of the issue, and at least two territorial governors, L. Bradford Prince and Edmund G. Ross, supported the creation of a judicial authority for settlement of land claims. Along with these men, Frank Springer provided the impetus for action. Springer traveled to Washington in 1890 and 1891 to lobby members of Congress and assist in drafting legislation creating a court of private land claims. He also gave Governor Prince extensive advice on the matter and urged that those supporting legislation on behalf of the territory of New Mexico adopt a unified position.

A court of private land claims was established by act of Congress on March 3, 1891. The newly created court was empowered to adjudicate claims arising in New Mexico, Arizona, Utah, Nevada, Colorado, and Wyoming, but the vast majority of cases heard involved New Mexico. Between its creation in 1891 and its termination in 1904, the court of private land claims considered 231 cases concerning claims for lands located in the territory of New Mexico. The court achieved partial success in resolving the confusion surrounding Spanish and Mexican land titles, but a great many claims were left unresolved.

MAXWELL LAND GRANT COMPANY V. JOHN B. DAWSON

Springer's most notable courtroom defeat came in a case initiated in 1891, just prior to his ascendancy to the office of president of the Maxwell Land Grant Company, and carried on for a period of several years following. The case originated as a suit of ejectment against John B. Dawson, who operated a ranch on the Vermejo River.

A native of Kentucky, Dawson had learned that there was money to be made in buying cattle and driving them to distant markets in California and Colorado. He invested in cattle in association with relatives, and later in partnership with Charles Goodnight and others. In 1868 Dawson learned of a parcel of land for sale on the Vermejo River. In partnership with Taylor Maulding, Joel Curtis, and Manly Chase, Dawson entered into the transaction, fronting the cash from a recent sale of cattle in Colorado.[6]

It was accepted as fact that Dawson had purchased land from Lucien Maxwell as early as 1868. Dawson held a deed from Maxwell and wife, dated January 17, 1869. When the dispute came to trial, Dawson and his attorney were not eager to introduce the document into evidence, for it was a most peculiar and confounding sort of instrument. According to this deed, Dawson had purchased from the Maxwells an indeterminate tract described as follows: "All the land or ground now suitable for farming or cultivating purposes in the valley or drainage of the Vermejo river, county of Mora, territory of New Mexico, within the following boundaries, to wit: Beginning at a certain dam at the head of a certain ditch at the right-hand point of rocks; from thence running down on the north side of said river to a certain other pile of rocks, on a knoll or elevation, with some bushes near thereto; thence running very near southward across said river to a pinon tree to the right of a ridge near a wash, which tree is marked with the letter 'L,' thence running up said river on the south side to the place of beginning, containing about _____ acres, more or less."[7]

The tract was presumed to be a part of lands excluded from the sale of the old Beaubien and Miranda grant when Maxwell sold the property in 1870. Maxwell estimated that such exclusions totaled approximately 15,000 acres. When he attempted to survey Dawson's land on behalf of the Maxwell Land Grant and Railway Company, an exasperated William R. Morley reportedly told Dawson, "I don't believe there is any man in the world can take that deed and by it locate any land."[8] Remarked William Keleher in retrospect, "This interesting description, certain to send cold chills up and down the spine of a title examiner, to cause the most resourceful pipe-smoking land surveyor to scratch his last match on a worn out pair of corduroy trousers, contained the germ of much litigation."[9] Notwithstanding the confounding nature of the deed, Morley had mapped out a tract of 1,260 acres, now conceded to be the property of John B. Dawson.

The late interest of the Maxwell Land Grant Company in asserting its right to lands used and claimed by Dawson is easily explained: the larger, disputed tract—an expanse of some 20,000 acres or more—was found to overlay a rich deposit of commercial grade coal. With this very large prize at stake, Springer could well afford to press the company's claim.

Finding himself the subject of a court summons, Dawson sought legal counsel capable of mounting an effective defense. Family and neighborly ties led him to Charles Springer, who had married Lottie Chase, the daughter of Dawson's longtime neighbor and business associate, Manly Chase. Charles Springer was described by historian Lansing Bloom as "a brilliant attorney who seldom practiced."[10] When Frank Springer moved to Las Vegas, his brother Charles remained in Cimarron, continued to build up the family ranch, and became involved in affairs of the Republican Party and the Territorial Stock Growers Association. Within the small legal fraternity in New Mexico Territory, it was common for attorneys to find themselves opposed in litigation by friends and colleagues. Such professional adversarial relationships rarely engendered ill feeling, and there is no indication that relations between Frank and Charles were strained as a result of the Dawson case.

Perhaps recognizing the perilous nature of his predicament, Dawson concluded an agreement with Charles Springer in which Dawson conveyed a one-half interest in the property in dispute "in consideration that the said Charles Springer should pay all costs and expenses of carrying on and conducting to a final determination the defence of said claims to said valuable lands set up and made by the Maxwell Land Grant Company."[11]

Charles Springer sought the services of an able young attorney, Andrieus Aristeius Jones, who later served as a U.S. senator for New Mexico. On October 10, 1891, Springer met Jones at his office in Las Vegas, and an agreement was reached whereby Jones would handle the matter for a consideration of five hundred dollars.[12] Several points of controversy later surfaced concerning the verbal contract for services, but Jones accepted the case and conducted a vigorous defense on Dawson's behalf.

Charles Springer. Courtesy of Museum of New Mexico, Neg. 7885.

Thomas B. Catron joined Frank Springer in prosecuting the claim of the Maxwell Land Grant Company. Both sides prepared their cases, and the matter came to trial in May 1892. Following a change of venue, the case was tried in Las Vegas, the county seat of San Miguel, before a jury composed mostly or entirely of Hispanic American citizens of the county.

On behalf of his client, Jones first claimed ownership of the larger tract on the basis of "adverse possession"—that is, practical possession and use of the property for a period of ten years or more prior to commencement of the suit. When the Maxwell deed was admitted into evidence over his objection, Jones shifted his strategy and endeavored to demonstrate that the property described in the deed was the same tract claimed by Dawson. When the case was appealed to the territorial supreme court, a dissenting jurist observed that Dawson sometimes claimed ownership under the deed and sometimes preferred to ignore it, claiming instead under adverse possession. This, the judge remarked, made for "an ingenious application of the doctrine, 'heads, I win; tails, you lose.'"[13]

In any event, the jury found in favor of the defendant, John B. Dawson. There is no reason to suppose that the jury made its decision on any basis other than its best judgment concerning the facts and the law; however, other factors are worthy of mention. First, it was not to be expected that a jury composed primarily of Hispanic men would exhibit a great deal of sympathy for the Maxwell Land Grant Company, which had ejected, among other settlers, a good number of Hispanic farmers and ranchers. Second, it is accepted tradition among descendants of Frank Springer and John B. Dawson that Dawson commanded immense respect among the people, and that no jury of ordinary citizens would have relished finding for a large, foreign-owned corporation in preference to Dawson.

Springer and Catron first appealed the case to the territorial supreme court, where it was heard in the summer of 1893. A majority of justices found the decision of the jury a reasonable one, and the rulings of the trial judge free of error. Springer immediately took the matter to the U.S. Supreme Court. The Court found in favor of the Maxwell Company and remanded the case to the territorial courts, instructing that the verdict be set aside and a new trial be granted.

Springer thus maintained what was at that time an unblemished record before the U.S. Supreme Court, but he could not prevail in his own backyard, as a second trial in the district court resulted in another verdict in Dawson's favor. Springer again sought relief in the territorial supreme court, but the appeal was dismissed and the matter was concluded with John B. Dawson secure in possession of some 24,000 acres, subject to the one-half interest conveyed to Charles Springer.

This ended the controversy, so far as Frank Springer and the Maxwell Land Grant Company were concerned, but a dispute later arose between

A. A. Jones and Charles Springer over the compensation awarded Jones for his successful defense of Dawson's interest. When it came to light that the Dawson Fuel Company was purchasing J. B. Dawson's coal-rich property for $450,000, and that Charles Springer was to receive half for his effort in securing a good lawyer to handle the ejectment suit, Jones sued seeking additional compensation. He had as yet received only the sum of $500 agreed on with Charles Springer, plus expenses incurred in traveling to Washington to argue before the Supreme Court. Whatever the merits of his claim, Jones did not succeed in prying additional compensation out of Charles Springer.

The Bent Claim

Springer argued at least eight cases before the U.S. Supreme Court, almost always as sole or senior counsel, losing only once. In the lone defeat, Springer sought to forestall enforcement of a lien placed against the Springer Ditch and associated lands by Patrick Ford, who had made substantial improvements to the ditch system under contract with the Springer Land Association; the Court found Ford's claims in order and affirmed the territorial court's decision.[14]

One of the most celebrated cases handled by Springer on behalf of the Maxwell Land Grant Company was the suit of the Bent heirs, who claimed an undivided one-twelfth interest in the grant. The controversy dated to September, 1859, when the three surviving children of the slain territorial governor Charles Bent brought a lawsuit claiming that by verbal agreement, Bent had obtained an interest in the Beaubien and Miranda Land Grant. In 1865 the Court determined that the Bent heirs held a one-fourth interest in the grant, which by this time was substantially owned by Lucien B. Maxwell. The Bent children indicated a willingness to convey their interests for a consideration, but before the negotiations were concluded, one of them, Alfred Bent, died. A consent agreement carried into effect what appeared to be the likely result of the negotiations: Lucien Maxwell was to pay each of Bent's surviving heirs six thousand dollars. A court decree to this effect was signed by or on behalf of Bent's two daughters and Guadalupe Bent, widow of Alfred Bent and the mother of his three children.

In connection with the sale of the grant to the Maxwell Land Grant and Railway Company in 1870, Maxwell and the company jointly brought suit against Guadalupe Bent—Guadalupe Thompson, following her marriage to George Thompson—to quiet any possible claim against the title. Mrs. Thompson contested the suit, claiming that language differences and ignorance of the law induced her to participate in the 1866 agreement against her children's best interests.

The case was in litigation for decades and twice came to the attention of the U.S. Supreme Court. Its final resolution came in an 1897 decision favoring the Maxwell Company. Two justices dissented, but the majority expressed itself through an opinion written by Justice David Brewer, then a member of the high court. The Court held that the disposition of the Bent claim in 1866 had provided reasonable compensation and was in the best interests of Alfred Bent's minor children. The Court further found that the transaction was consistent with Bent's intent had he lived, and that the transfer of ownership had been legally executed.[15] Reporting the result, the *Las Vegas Optic* acknowledged Springer's triumph and said of the suit, "It is the last of a long list of cases involving the title to the Maxwell grant."[16]

Professional Standing

Of such victories was Frank Springer's reputation made. His colleagues took note not only of the favorable results of his pleadings, but the methods by which they were gained. Of Springer's capacity for finding the essence of a controversy, Ralph Emerson Twitchell recalled that he had "a peculiar facility for adjusting disputes, differences in matters of business, and in affairs affecting the community. More than one instance could be recalled where after listening to a whole evening's wrangling, with no prospect of a conclusion, he has offered a solution in a brief paragraph, accurately stated, that was accepted by all concerned as satisfactory." In legal practice, this capacity was evident in the reputation acquired in later years, "of making short arguments in large cases."[17]

Springer was well regarded among judges, as among his peers in the legal profession. Springer valued the confidence of the judges before whom he practiced. His respect was generally reciprocated by the judges, who presumably were among his best-informed critics. Judge J. W. Crumpacker once offered an impression of the Springer style based on a case carried by Springer to the territorial supreme court. It was, as Crumpacker recalled, "the first time I had appeared before him." The case was a complicated one involving the application of a new law governing water rights. As related secondhand by Twitchell:

> Several points had figured in the trial of the case in the lower court, which were discussed by opposing counsel at length. Springer took them up in order when he came to reply, and after stating the first one said he was uncertain about it; as to a second one he wasn't quite sure; and when he continued to profess ignorance about a third, [Crumpacker] said to himself, "What *does* this fellow know, anyhow?" Upon the next proposition, however, he declared that he was absolutely certain; and this he proceeded to demonstrate, using a blackboard which he had brought into court, in such a way that the Court found itself certain too. "And then we discovered," [Crumpacker] said, "that the only point which this ignoramus knew was the one upon which the whole case turned, and that it didn't make any difference which way the others went; also that the way he knew it was that which necessarily led to a decision in his favor; and I was instructed to write the unanimous opinion of the Court accordingly."[18]

Much of Springer's legal work was done on behalf of the Maxwell Company, the Atchison, Topeka and Santa Fe Railroad, and other corporate clients. He did, however, represent some notable individuals in the course of his legal career. Springer was an admirer of Robert G. Ingersoll, an attorney and orator of note, a Republican Party politician, and a strident agnostic. Having represented Stephen W. Dorsey, a defendant in the "Star Route" fraud cases, Ingersoll was due substantial compensation. Unable to pay his debt immediately, Dorsey strung Ingersoll along, promising payment as he was able to liquidate cattle and other assets of the Palo Blanco Cattle Company. As a courtesy, Springer acted as Ingersoll's collection agent between 1889 and 1892, monitoring Dorsey's transactions and securing payment of sums owed to Ingersoll.

In 1895 Springer was one of four prominent attorneys representing Thomas B. Catron, who was threatened with disbarment in connection with alleged witness tampering. Some believed that the charges were politically motivated, and the circumstances provided a good deal of evidence to that effect.[19] The matter went to trial, and Catron and his law partner, Charles Spiess, were found not guilty of the alleged offenses. Immediately thereafter Catron was elected president of the New Mexico

Thomas Benton Catron. Courtesy of Museum of New Mexico, Neg. 50616.

Bar Association, and he went on to serve as one of New Mexico's first two U.S. senators.

Springer's active legal career continued until 1906, when a physical breakdown was the proximate motivation for his retirement from the regular practice of law. Although he recovered and enjoyed many more productive years, his legal activities were confined to occasional efforts on behalf of the Maxwell Company and the cultural organizations with which he was affiliated.

Reflecting on Springer's career at law, a longtime friend and colleague, W. C. Reid, remarked, "What a wonderful lawyer he was. Law is not an exact science or a science at all, but in his hands it almost became so. In treatment of a case in his brief or argument, to the Court, there was never a word too much nor words too few. There were no technicalities about his arguments. He was too great a man to think of small things. His conclusions were reached by reasoning so clear that there could be no mistake."[20]

If, following his retirement from the active practice of law, Springer missed the intellectual stimulation of legal research and argument, the fellowship of the bar, or the theater of the courtroom, he gave no such indication. His legal career had been a means to an end, and he had too much to do and too much to look forward to in other lines to concern himself with the past. Having achieved his goal of financial independence and having distinguished himself in the process, he turned his attention to the occupation that had interested him from his college days, and on which he could now concentrate his full energies—the life of a working scientist.

CHAPTER 9

The Normal

Once settled in New Mexico, Frank Springer assumed the role of an engaged and constructive citizen and became a champion for education. In the course of his career as a public figure, he frequently spoke on the philosophy and practice of education, which he regarded as the foundation of civilized society. In his two terms as a member of the territorial council—separated by an interval of more than twenty years—his primary interest was in legislation supportive of a strong system of public education.

After some fourteen years in Las Vegas, Springer was able to be of service in the development of higher learning as a regent of the new Las Vegas normal school. The institution had been awarded to Las Vegas some years previous but had been unable to open for want of facilities and the direction that a unified board might have provided. The challenge of bringing a new institution into existence, establishing its founding principles, and influencing its development was one that Springer embraced, and for which he was singularly well prepared.

In 1893 the territorial legislature had passed, and Gov. L. Bradford Prince had signed, legislation establishing two normal schools, or teachers' colleges—one to be located at Silver City in the southern part of the territory, the other at Las Vegas to serve the northern counties. The implementation of this well-intended public act was attended with repeated delays and difficulties. In the case of the Las Vegas school, regents were appointed and the new board met and organized itself in March 1893.[1] But when it came time to take specific steps to bring the school into being, controversy ensued. Most at issue was the matter of location. Between the private allegiances of individual board members and the mutual jealousies of New Town and Old Town, various sites were proposed, virtually all of them unsatisfactory to one faction or the other. The legislature finally settled the matter in 1895, amending the Normal School Act to specify a location that lay in East Las Vegas, but that was convenient to residents of the older west side as well. The legislature also authorized the issuance of thirty-five thousand dollars in bonds for public projects, including ten thousand dollars to build and equip the Las Vegas normal school.

By 1897 the school had existed on paper for four years, the board had met regularly and regents had come and gone, yet the commencement of a viable normal school operation was not in view. However, events were converging to move the school forward, and the appointment of Frank Springer to the board of regents in the spring of 1897 proved to be a critical turning point. To the responsibility of stewardship for a new institution of higher learning, Springer brought his expertise as an attorney, the credentials of a respected scholar and scientist, rhetorical skills of a high order, and business acumen. Most of all, Springer brought a vision of the possible that transcended conventional limits and paid no heed to the lamented conditions of poverty and isolation so often used to excuse New Mexico from higher achievements in its public endeavors.

At a meeting held on April 15, 1897, Springer was recognized as the board's newest member and was immediately elected chairman. As the school had no campus, it was agreed that meetings would be held in a room on the second floor of the wholesale mercantile firm, Browne and Manzanares Co., the place of business of Millard W. Browne, secretary-treasurer of the regents.[2] Charles Ilfeld, Charles Rudolph, and W. S. Burke were the other members of the board.

By April 27 Springer had become president, rather than chairman, of the normal school board, although minutes reflect no official action specifying the change of title. A called meeting of that date dealt with the sale of bonds for construction. Minutes of the next several meetings reflect a continuing concern with construction and financial matters related to the school's opening. Equally pressing was the need for capable leadership, in the person of an administrator with the creativity and determination to assemble a strong faculty and implement a rigorous academic program.

The Young Educator

At about this time Springer's interest in evidence of prehistoric life in New Mexico brought him in contact with a young professor at the Colorado Normal College at Greeley. Just thirty-one years of age, Edgar Lee Hewett held the responsible position of superintendent of teacher training at Greeley.

Hewett was also interested in anthropology and was knowledgeable concerning the ruins in Frijoles Canyon and the ancient culture to which they bore witness. Springer heard Hewett speak and was impressed with his familiarity with antiquities that were near at hand but mysterious to non-native residents of the Rio Grande Valley and nearby communities across the mountains. At Springer's urging, Hewett was invited to address a gathering of citizens in Las Vegas. A short time later, with the apparent blessing of the other regents, Springer invited Hewett to Las Vegas to discuss the possibility of his heading the new normal school.

Hewett was interviewed by Frank Springer and M. W. Browne. Upon their recommendation, the board voted to employ Hewett as principal of the normal school. The starting date was indefinite, but it was agreed that he would report well in advance of the school's opening in order to hire a faculty and prepare for the school's first students.[3] Hewett later received a contract naming him president, rather than principal, of the new institution.

"My presidency at the normal university was an exciting adventure in education," Hewett recalled. "I was pretty young for an executive, but I had a year in which to prepare for it, in which I studied my problem intensively." In Hewett's view, the essential problem was to design an institution for teacher training that was "adapted exactly to the needs of the population. . . . I created that in my mind," he recalled, "and then manned it with about the finest young faculty ever assembled." By June 18, Hewett had relocated to Las Vegas, and was busy assembling a faculty and making other preparations for the school's opening.[4]

Institution Building

When Hewett reported for duty, much remained to be done. Though many months under construction, the normal school building remained unfinished, and the funds appropriated for construction were insufficient to finish the job. Under the leadership of Springer and Hewett, resources were found in the form of a subscription organized among local businessmen.

In the Las Vegas school's approach to construction of facilities, Springer found a defining value of the institution. Springer took no particular credit for the virtue he pointed to, but acknowledged the regents who had preceded him on the board, commending them for their vision in building well instead of fast. "They planned not so much for the immediate needs as for something that would be a permanent monument, an ornament to the City, and worthy of the cause this school was intended to promote—an inspiration and a delight for future time." By way of contrast, Springer could point to the other state normal school created in the same legislation by which the Las Vegas school was authorized. That institution, he said, "seeking a quicker result, builded for the present with what means they had, so that they had their school in operation, when we had only one story of an uncovered building to show."[5]

Undaunted, Hewett advertised the opening of the new normal school, advising all comers that the school would open on the first Monday in October 1898. More than ninety students enrolled for the first term. The inconvenience that they experienced as they attended classes on the first floor while work proceeded on the upper floors may have been offset by the realization that they were part of an adventure in educational innovation. By February 1899, the upper floors had been completed and the first floor was turned over to the institution's "model school." Built of purple stone quarried and cut near Las Vegas, the normal school building was an impressive structure, rising from a broad knoll at the northern extent of the city and overlooking the commercial district of East Las Vegas. Romanesque in style, the building was the visible symbol of a pervasive commitment to excellence.

At the heart of the university was the Normal Department, established for the training of teachers. The Academic Department provided a four-year high school curriculum and preparation for college studies. The Preparatory Department provided for students who needed remediation or review preparatory to attempting postsecondary work or sitting for teachers' examinations. A graduate school provided two years' study beyond the basic teacher preparation course; successful completion resulted in conferral of the master of pedagogy degree. At various times the institution included a commercial school, a manual training school, and model schools for all grade levels.

These were conventional activities for a state normal school. But Hewett's vision—and Springer's—went beyond then-accepted practices for normal schools. Hewett believed that prospective teachers could benefit from participation in original research, as did graduate students training in universities. So enthusiastic was Springer with the vision on which the school was founded that he prevailed upon the territorial legislature in

Main building, renamed Springer Hall in 1919, New Mexico Normal University. Courtesy of Citizens Committee for Historic Preservation, Las Vegas, New Mexico, Neg. 0043.

early 1899 to change the name of the institution from the "New Mexico Normal School at Las Vegas" to "New Mexico Normal University."[6]

The building was completed in early spring and dedicated on March 4, 1899. Before the assembled students, faculty members, townspeople, and visiting dignitaries, Springer declared the new institution "a school of the people and for the people." Of the institution's core purpose, he declared, "The teachers who go out from here are to be the apostles of that modern system of education which holds that none of the powers and faculties of human beings are unworthy of its care; which gives attention to muscle as well as to mind, and to hand as well as to brain."[7]

In its first full year of operation the Normal enrolled 195 students and sent forth its first graduates. In 1899–1900 enrollment rose to 242, and the following year to a total of 308 students.[8] Among those enrolled in the early years of the normal university were at least four of Frank Springer's children, Eva, Ada, Edward, and Henry, Eva graduating with the class of 1901 and Ada in 1902.[9]

In the new institution there was abundant evidence of the faculty's creativity. The university operated a biological field station and carried on archeological investigations in the Pajarito region. By December 1898, the school was publishing a literary journal, the *Southwest.* An oratorio society organized under auspices of the university at times boasted one hundred voices and undertook performance of Haydn's *Creation* and comparably ambitious works.[10] Literary societies gave students opportunities to practice parliamentary procedure, to gain skills in essay writing, debate, and oratory, and to develop "ease and composure of manner when appearing before an audience."[11] By fall of 1899, the school had added instruction in fine art, employing Kenneth Chapman, a Chicago-trained artist, as instructor.

Springer of San Miguel

Perhaps as a result of his involvement with educational issues, Frank Springer sought elective office for the third and final time in the fall of 1900, winning the nomination of his party to represent San Miguel County in the proximate biennial session of the territorial council. Springer was suggested as a candidate for high office on other occasions, but he was not inclined to promote himself as a candidate for election or appointment. He was an active member of the Republican Party, but he sought elective office only on his own terms and for his own reasons.

The election year 1900 found New Mexico's Republican Party split into contentious factions, with Thomas B. Catron leading the old guard and Gov. Miguel Otero representing the insurgents. In the county nominating convention, Springer—who was aligned with the Catron wing—prevailed over his Republican opponent and won the party's nomination for one of two council seats allotted to San Miguel. This result pleased the *Optic,* which, however, did not think much of the Republicans' other choice, Charles Spiess.[12]

In the general election Springer faced the Democratic nominee, Charles Tamme. Although it was uncharacteristic of the *Optic* to oppose Springer in anything, the paper endorsed the Democratic ticket in its entirety. The *Optic* offered no further comment on Springer's candidacy, but heaped scorn on the head of Spiess, who was characterized as a turncoat, an ingrate, an opportunist, and a chronic office seeker.[13]

Nonetheless, in its edition of November 7, the *Optic* reported a "sweeping victory for the Republicans" and published unofficial returns showing both local candidates for the council with substantial margins of victory over their Democratic opponents. The writer attempted to put the best face on the election, counting no fewer than seven members of the new council who could be relied on to oppose the governor. Of Springer, the *Optic* said, "Frank Springer, of San Miguel County, is generally thought to be too big a man to lend himself to participation in any nefarious coal oil, jury law or salary-grabbing schemes, and will likely be found voting with the foregoing named on important matters." Four others, including Spiess, the paper declared, would have to "flock by themselves."[14]

When the legislature convened in January 1901, Springer was named chairman of the council's Education and Judiciary committees. He took an active role in shaping legislation concerning administration of the courts and rules of evidence, but his primary mission was the enactment of a comprehensive public school law designated as Council Bill 23. The session opened with a centennial observance commemorating the appointment of John Marshall to the U.S. Supreme Court in 1801. Springer delivered the keynote address and the council responded with a resolution of thanks for "a very eloquent, learned and interesting address."[15]

The "Springer School Bill" proposed to increase the size of the territorial board of education and specified more rigorous procedures for the examination and licensing of teachers. Three classes of teaching certificates were created, reflecting varying levels of preparation. The bill also divided school districts into three classes and specified provisions for the election of local school boards.

The proposed legislation reflected model practices for the administration of public schools, as viewed by Springer and Hewett. Said the *Optic,* "A careful reading of Council Bill 23, shows it to be exactly such a bill as the people would expect Mr. Springer to introduce. It bears no marks of subservience to the political or selfish ends of any person or class. It is evidently drawn with a sincere desire to benefit all the people by improving the public schools." The paper found special merit in provisions intended to insulate school boards from political manipulation and to raise the qualifications of teachers. The writer concluded, "There is not a single provision of this bill that is objectionable to fair-minded people. If it is opposed at all, it will be by those who have selfish interests to look out for and who care nothing for the welfare of the schools."[16]

Springer's education bill passed the council but encountered opposition in the House. The *Las Vegas Optic* perceived a single threat to the Springer bill, in the person of Frank A. Hubbell, easy boss of the Bernalillo County

Republican Party and chairman of the territorial Republican Central Committee. Along with Solomon Luna, a member of the national Republican Central Committee, Hubbell was said to have been making the rounds in the legislature, doing the governor's bidding and making mischief for the school bill. The *Optic* denounced Hubbell as a man of sinister motive and an obstructionist to education, and accused him of threatening members of the legislature with defeat of their bills if they did not accede to his wishes and vote against the school bill.[17] Despite this, the Springer bill passed both the council and the House with minor amendments and was signed into law by the governor.

With the passage of the school bill, Springer saw that model practices were incorporated in official policy for the administration of public schools in the territory. As was frequently the case in his mature years, Springer also managed to retain the respect and good will of almost everyone concerned, while taking part in what was essentially an adversarial process.

An Honorable Retreat

Under the leadership of Springer and Hewett the university continued to grow, and its reputation grew with it. By January 1903, A. E. Winship, editor of the *New England Journal of Education,* could declare the New Mexico Normal University "one of the best normal schools in the country." Winship found the school notable for exemplary programs in biology, anthropology, and art, and said of the program in the natural sciences, "I have never seen it excelled." He was impressed with the physical plant and pronounced it "one of the best normal school plants of its size in the country."[18] This was heady praise for a young institution in an impoverished western territory, but by this time there was trouble in paradise, as a board Springer had led in substantial harmony splintered into sharply divided factions.

Tension had been building for some time, but the trouble began in earnest when, in mid-October, 1902, George Selby, a longtime member of the board of regents, announced his resignation, explaining that he was soon to leave the territory.[19] In his place, Gov. Miguel Otero appointed W. R. Tipton, a Las Vegas physician who was determined to reverse policies championed by Hewett and Springer. Tipton presented a letter of appointment and was seated as a new regent at the board's meeting of November 13.[20] Whether Otero intended to shake up the Normal by means of Tipton's appointment is not clear. Subsequent events did, however, reveal that the governor was in support of the dissident regents who were now in the majority.

Whatever may have taken place between Hewett and the hostile regents, no record of overt conflict made it into the regents' minutes, nor was the controversy initially reported in the *Daily Optic*. Not until January 15, when Hewett tendered his resignation, did the affair become a matter of public concern. On that date, the board held its regular meeting. Hewett's resignation was offered without fanfare or discussion at the end of the meeting. Hewett noted that his contract would "expire by limitation" on June 30, 1903, and stated that he would not seek reappointment.[21] Hewett's simple announcement set off a more tumultuous chain of events.

Four days after Hewett's announcement, Frank Springer called a special meeting of the board to announce his resignation, effective immediately. As reported in the minutes, the other regents received the news with regret and responded in a conciliatory manner. "Every member present expressed regret that Mr. Springer would sever his connection with the board, and he was requested to withdraw his resignation, which after some discussion he declined to do."[22]

Springer's resignation prompted a barrage of letters and petitions directed to Governor Otero. On January 20 pioneer wholesaler and fellow regent Charles Ilfeld wrote to urge that Springer's resignation be refused.[23] On January 21 the faculty issued a statement naming Tipton as the source of the school's difficulties and urging that he not be reappointed at the expiration of the term he was completing in place of the departed regent Selby.

Otero wrote Springer to ask that he remain on the board until February 22, when appointments for Springer's seat and the seat held by Tipton could be sent to the territorial council for confirmation.[24] Springer's response left little doubt that his decision was firm, as he told the governor, "It will be satisfactory for me to continue to act with the Board, if there should be any occasion for it, until the time you mention, when I shall be pleased to be relieved."[25]

More than sixty Las Vegas citizens signed a petition dated January 28, 1903, expressing concern over Springer's resignation and telling the governor: "Mr. Springer's interest in the success of the University, and his efforts therefore, have been invaluable, and we do not believe his place can be filled. He has contributed legal and official services to the Board and University, and although the law provides compensation, he has never accepted it, but as a public-spirited citizen he has made gratuitous contribution of his time and talents to the betterment of the school since its inception. We believe it will be a calamity to this community if you accept his resignation, and we therefore respectfully petition that you refuse to accept the same."[26] Another petition in support of Hewett and Springer was forwarded from an organization of students of the Normal.

By January 29 the *Optic* could no longer ignore the controversy, but the paper ventured only a tepid call for local citizens to stand by the Normal's leaders, "uphold the hands of those connected with them, and frown down any effort at unjust and unreasonable criticism." On January 30, the board of regents, minus Springer, held a somewhat contentious meeting, at which it was reported that the students were "holding mass meetings and passing or were about to pass resolutions derogatory to this board or to some of its members, and other matters which were a subversion of discipline and against the dignity of the board." Hewett was not present during the initial discussion, but was "called in and questioned on the subject." Hewett disavowed knowledge of much that was alleged, but he did acknowledge that a student meeting was scheduled for that afternoon. It was then ordered, on a motion adopted by the board, "that the faculty should censor any resolutions offered at the students['] meeting and not permit any disrespectful or undignified resolutions to be issued by said students." Hewett was further requested to respond to a rumor that the faculty was planning to resign en masse. Following a consultation with the Executive Committee of the faculty, Hewett reported that the rumor was untrue, "but further stated that it was the intention of the faculty to resign at the expiration of the present year."[27]

The following day the *Optic* weighed in with greater force in an editorial titled simply, "Frank Springer." Said the paper, "The *Optic* in common with very nearly all the citizens of Las Vegas regrets the resignation of [the] Honorable Frank Springer from the board of regents of the Normal University. It hopes that the cause which induced this act may be removed, and that Mr. Springer may see his way clear to again take his place on the board of regents. He has been connected with the Normal school from the beginning and has rendered such valuable service that the public cannot afford to lose him."[28]

On February 10, Governor Otero again wrote to Springer, acknowledging his letter and the note reiterating his desire to resign his seat on the board of regents, adding that "it would give me great pleasure to have you withdraw your resignation if you should see fit to do so, especially as the residents of Las Vegas desire that you remain on the Board."[29] Otero well knew that a change of heart on Springer's part was unlikely.

Several possible causes of the controversy have been suggested. Some believed that the root cause was a conflict over use of the university's athletic field for Sunday games. Another bone of contention was the university's conduct of a secondary school program under contract with the local school board. Other citizens may have been incited to anger over a rumor that Hewett had traveled to Washington to oppose statehood, and still others objected to his role as an advocate for preservation of public lands. Lynn Perrigo has further surmised that Hewett "was under fire for promoting research on remote topics instead of orienting the institution directly to the culture and problems of the region."[30]

An additional cause of dissatisfaction with the normal university administration, perhaps the decisive one, was Hewett himself. Despite his many achievements and the admiration of loyal supporters, Hewett sometimes displayed an abrasive manner that engendered strong resentment. Over nearly a half century of public life in New Mexico, Hewett frequently found himself embroiled in conflict; as often as not, he became the issue, inciting hostility on the part of his critics and turning his friends into defenders. According to Kenneth Chapman, Hewett's problem in Las Vegas was compounded by the perception that he held himself aloof from the community, keeping to a well-worn path between the campus and his home, spending much of his time in solitary research, and spending school vacations away from the city.[31] It is noteworthy that Governor Otero and the regents, in apparent good faith, encouraged Springer to reconsider his resignation and remain on the board. Neither the governor nor the new majority of regents had any intention of retaining Hewett.

If the issues appeared somewhat muddled, it was not the fault of the faculty, which on February 11 issued an unequivocal declaration stating that the school's troubles could be traced to the appointment of a regent "who is well known to be an enemy of the management of the school, avowedly hostile to its methods and aims."[32] Tipton was not mentioned by name, nor was such mention necessary. The faculty statement enumerated what the writers believed to be the motives of the school's critics:

> Inability to dictate the appointment of some friend to a place on the Faculty.
>
> Inability to dictate the removal of some teacher.

Inability to secure a place on the Faculty.
Inability to secure certain business patronage.
Resentment of correction or discipline of pupils by the Faculty.
Resentment against the stand of the school against the use of its athletic field for Sunday sports.[33]

Theo Cockerell, a distinguished biologist, alerted his profession to the situation through a note published in the journal *Science,* identifying political interference as the root cause of the turmoil. The school, Cockerell wrote, had previously enjoyed "a most fortunate immunity" to such interference, thanks in large measure to "the wisdom and influence of Mr. Frank Springer, the well-known authority on crinoids, who has been president of the board of regents."[34]

The *Optic* belatedly took hold of the issue, editorializing extensively on February 16 and returning to the issue in editorials published on February 19 and 20, but the die was already cast. On February 23, Governor Otero appointed W. R. Tipton to a new five-year term on the board of regents and chose Charles Spiess of Las Vegas to fill the seat vacated by Springer. The spring term ran its course without further incident, but on commencement day, May 28, all thirteen members of the faculty signed a letter of resignation that was forwarded to the regents.

With this experience behind him, Hewett set about the business of obtaining the credential that would legitimize his career as a scientist, admit him to the exclusive company of learned societies, and open doors to other opportunities as inviting as the one he had been offered in Las Vegas. Within a few months Hewett had been accepted for doctoral study at the University of Geneva and had sailed for Europe.

In retrospect, Hewett could view the events of 1902–1903 philosophically. For him, the decisive factor in the demise of his administration was the displeasure of a governor who "had the whip hand" by virtue of his power to appoint regents. In Hewett's view, the governor saw the university as "too much of a happy family affair"—meaning, apparently, that faculty appointments were determined by the administration and faculty, and were not subject to political manipulation. Of Otero, Hewett observed, "He was one of the few men I ever knew who could withstand the logic of that great friend of education, science, art, and everything else that was good for his state, whom I have always looked upon as the Pericles of New Mexico, Frank Springer."[35]

Whatever the immediate cause of Hewett's ouster and the resignation of Frank Springer from the board of regents, the affair revealed significant division in the community over the merits of Hewett's leadership. While the administration of the Normal under Springer and Hewett can be admired for its energy, creativity, and commitment to quality, it was not as responsive to needs of the region as it might have been. For some local residents, the activities that Hewett promoted, including scientific and oratorio societies, archeological field research, and the publication of a literary magazine, may have had slight relation to the legislative mandate creating a "normal school" for the express purpose of training teachers.

While a majority of those in the region were of Hispanic origin, few Hispanic students were enrolled in the Normal in its earliest years and no Hispanic faculty members were employed. Neither was there a noticeable increase in Hispanic participation in the decades following Hewett's departure. Not until the early 1970s, following a change of name to New Mexico Highlands University in 1941, was there an explicit reorientation of the school's mission to recognize a primary obligation to the region and its large Hispanic population.

Despite the unhappy circumstances, Springer's withdrawal from active service as a regent was not a bitter parting. To the Normal Springer had given freely of his time, his wealth, his influence, and his ample vision. He never thought of building an inferior or average institution, nor was his vision constrained by the territory's meager financial means. If a forced change in the direction of the Normal was, to some extent, a rejection of some of the notions championed by Springer and Hewett, that fact did not suffice to alienate his affection or extinguish his enthusiasm for the institution. Despite later efforts to return him to the board, Springer would not again serve as a regent, but he remained a loyal, constructive, and influential supporter of the Normal.

Neither did the normal school experience mark the end of Springer's association with Edgar Lee Hewett. The relationship that had been so productive throughout the formative years of the normal university, of the visionary enabler and the skilled implementer, was later to be resumed in another context.

CHAPTER 10

A Bend in the Road

By the time Frank Springer assumed the presidency of the Maxwell Land Grant Company, the grant had been under corporate management for more than twenty years. The property had been sold for taxes, had been the subject of foreclosure suits, and had twice been forced into receivership. The company had been reorganized at least twice and had been plagued with a succession of inept and corrupt administrative regimes. It had survived a bitter struggle over its title to the property, and more recently, an internal power struggle among the trustees. Endowed with abundant resources, the grant had been plundered by opportunists, but there had been little productive activity to justify the hopes of the company's investors.

Springer's ambition and his duty as president of the Maxwell Company converged in the conviction that timely, systematic, and well-managed exploitation of the grant's resources was of paramount importance. Upon assuming the presidency in 1892, he had intensified efforts to maximize revenues through increased land sales, town site development, livestock production, timber sales, irrigation, and exploitation of mineral resources—chiefly gold and coal. No possible avenue was neglected, but Springer had become convinced that land sales, livestock operations, and related enterprises would never suffice to put the company on a strong financial basis.

In the view of Lawrence R. Murphy, Springer was chiefly responsible for policies that gave rise to a revival of mining in the Baldy Mountain region. After an initial burst of activity following the discovery of gold in 1866, mining had gone into a steep decline. By the time Springer assumed leadership of the trustees, production had all but ceased and once-lively camps were nearly deserted. According to Murphy, "Springer was well aware of the difficulties and proceeded with a well-calculated program to attract miners and capital to the district."[1]

New regulations were adopted in the spring of 1894, with the evident intent of attracting both large investors and prospectors. Significant capital was needed to support large-scale mining involving the development of shafts, transport facilities, and mills to crush the ore and extract the gold. At the same time, it was desirable to provide incentives for the small-time prospector to do the initial work of locating promising ore bodies. By 1897 the company was well pleased with the increase in activity and so advertised in its "Mineral Regulations for Prospectors and Intending Locators," advising that more than five hundred claims had been located and 130 patents issued under the regulations.[2] Sixty locations had been staked in the vicinity of Baldy Town; twelve mines were producing ore; and four processing mills were in operation.[3]

In a notable experiment on the Moreno Valley side of Baldy, a group of Chicago businessmen formed the Oro Dredging Company and built a massive dredge christened the "Eleanor." In spite of early successes, the company went bankrupt. By the end of 1905, operations had been suspended. Charles Springer purchased the abandoned dredge, but was unable to resume operation of the Eleanor.

Developing the Maxwell Grant

It had long been clear to Springer that the company's best and perhaps only chance for financial viability lay in exploitation of its coal resources. Of the 1,714,764 acres surveyed and confirmed to the Maxwell Land Grant Company, half a million acres were found to overlay significant deposits of coal. In some places the coal was easily accessible and was of sufficient quality for processing as coke. Coal had been mined commercially since at least 1880, when the Raton Coal and Coke Company was established by the Maxwell Company in collaboration with the Santa Fe Railroad. The Santa Fe provided rail access to the mine sites in York Canyon and Dillon Canyon near Raton, while the Maxwell Company leased coal-bearing lands to the new company at terms agreeable to both parties.

Upon assuming the presidency of the Maxwell Land Grant Company, Frank Springer made an immediate push for investment in coal, publishing a detailed treatise on the subject for the benefit of the American trustees and the Dutch bondholders.[4] He cited increases in U.S. coal

Headquarters of the Maxwell Land Grant Company, Raton, New Mexico. Courtesy of Center for Southwest Research, General Library, University of New Mexico, Neg. 000-147-0096.

production since 1820 and demonstrated that there had been a dramatic increase in demand. He argued that with the rapid development of the western United States, there would be a growing market for coal. Springer believed that through a bold investment, some company—perhaps the Maxwell—could secure a virtual monopoly on coal in the Southwest, as others had done in the eastern states. He offered in evidence a detailed analysis of the geology of the area prepared by his friend of Iowa City and Burlington days, Orestes St. John.

As always, the impediment to development of new mines was transportation. Springer had analyzed distances, routes, and connections, and he laid out options for the construction of track to carry the Maxwell coal to market. Although the Dutch bondholders had previously declined to invest in building a railroad, Springer was vehemently opposed to the notion of involving an established carrier in transporting coal from new mines. "I am unwilling to consider the question of turning over these magnificent coal interests to the tender mercies of railroad companies, over which we have no control," he wrote, "until I am finally assured that there is no possibility of anything better, and that the friends of Maxwell in Holland would rather give away their most valuable assets than to invest the money to make them profitable."[5] The memorandum bore the signs of Springer's usual thorough preparation and impeccable logic. It did not, however, move the wary Dutch bondholders.

At Springer's suggestion, the Maxwell Land Grant Company assumed full responsibility for management of the mining operation of the Raton Coal and Coke Company in 1897, with the company's nearby Raton office providing an efficient base. The Santa Fe Railway retained a half-interest in the company, but in lieu of a share in profits generated, the railroad would purchase coal from the company at an agreed-upon rate and would receive a royalty of five cents on each ton of coal sold.[6] Springer now assumed responsibility for leading the coal and coke company, along with his duties as president of the Maxwell trustees. In recognition of his increased responsibilities, Springer was awarded an annual salary increase in the amount of fourteen hundred dollars.

The collaborative arrangement with the Santa Fe had worked reasonably well for nearly twenty years and had provided a solution to the Maxwell Company's problem of raising capital to build its own transportation system. The interests of the two companies, however, did not always coincide. As coal deposits at the earliest mine sites became depleted and Maxwell officials considered possible new locations for mining, a divergence of interests was apparent. Riding a wave of increasing demand and eager to gain on its financial obligations, the Maxwell Company wanted to develop mines of its own on lands not included in the original Raton Coal and Coke Company lease. At the same time, the Santa Fe was not eager to extend rail service to new mines that could only compete with those in which it held a financial interest. An accommodation was reached when the Santa Fe agreed to exchange its interest in the Raton Coal and Coke Company for additional acreage in York Canyon and build a branch line to Willow Creek Canyon, where the Maxwell officials contemplated opening new mines.[7]

Free from the entanglement with the Santa Fe, Springer and the Maxwell Land Grant Company could more aggressively develop the grant's coal resources, with only its own interests to consider. The company took a major step in this direction in 1901, selling the Colorado portion of the grant to Colorado Fuel and Iron for the sum of eight hundred thousand dollars. This sale enabled the company to pay off a loan of fifty thousand dollars previously advanced by the Dutch and settle obligations arising from termination of the Maxwell Cattle Company and Frank Sherwin's administration.

Beyond these steps, the use of proceeds from the Colorado sale was a matter of some contention. To the consternation of some investors in Europe, the American trustees did not use the funds to retire old bonds, but instead brought interest obligations current and refinanced much of the old debt, providing the trustees substantial cash with which to further develop the property. The trustees managed to mollify the Dutch bondholders, but controversy arose when investors holding some of the company's bonds in Scotland threw up a loud protest through the Edinburgh firm of Gill and Pringle.

In response to the complaints of the Scottish investors, holders of approximately 6 percent of the prior lien bonds, Springer wrote to Wertheim and Gompertz, the Dutch firm representing Maxwell bondholders in the Netherlands. "As to the general complaint they make, that proceeds of sales have been employed in the development of the property, instead of being used to retire bonds," he wrote, "that is a matter of policy in which we had reason to suppose we had the support of the Administrative office [in the Netherlands], representing the interest of 9/10 of the bondholders. It is perfectly clear that the only way to ensure any considerable revenue out of the Maxwell property is by the development of its coal mines. It is also the only way to build up a value which will render the property salable at adequate prices. It has been my idea, as you know from former correspondence, to get the coal interests into shape and try and sell them for a figure which you would consider proper."[8]

Springer expressed confidence that mines under development in Willow Creek Canyon would enhance the property's value significantly and pointed out that investments in development had revealed the existence of a coalfield of immense potential. As to the suggestion of the dissident Scottish investors, calling for appointment of a receiver to assume management of the company's affairs in New Mexico, Springer advised that court action would not be necessary to bring about a change in management. "I am ready, at any moment, to give way if it seems evident to you that better results could be secured through another," Springer wrote. "But I do not consider it proper or necessary that this Board, or ninety percent of the bondholders, shall be obliged to submit to the dictation of ten percent, as to these matters of policy which relate to the best interests of all concerned."[9]

As usual, the Dutch directors supported Springer, making interest payments to the Scottish bondholders on the same basis as to the Dutch, but turning back the notion that a receiver should be appointed to supersede the trustees. A conciliatory letter was written to mollify the Scottish investors. Springer was so advised by Wertheim and Gompertz who, at the same time, assured him that the Dutch had no intention of applying to the court for a change in management. With reference to its missive to Edinburgh, the Dutch firm noted, "You will see from the enclosed copy that we were rather hard on your Board, but you will understand that such is to be taken *cum grano salis*."[10] The words of reproach for the American trustees were written for the benefit of angry minority bondholders in Scotland. Springer took them with a grain of salt and continued in pursuit of what he believed to be the best interests of the trustees and bondholders.

As to the sort of major investment needed to develop a coalfield, the Dutch were no more willing at the turn of the century to sink additional funds into the property than they had been in 1874. Having accepted the reluctance of the Dutch to invest in transport facilities, plant, and equipment, Springer looked elsewhere for needed capital. In 1902 a group of Saint Louis businessmen was sufficiently interested to organize the New Mexico and Pacific Railroad Company and issue a prospectus for the information of potential investors. The company proposed to lay track from Raton to the mouth of Cimarron Canyon—a distance of some forty-seven miles. From Cimarron, the line would run up the canyon and into the Moreno Valley, terminating at Elizabethtown.[11]

Delays in financing derailed the effort, but by 1905 the organizers were ready to move forward, reconstituted as the St. Louis, Rocky Mountain and Pacific Company. In a letter to Fisk and Robinson—a New York firm engaged to market the company's bonds—Henry Koehler Jr., president of the new company, reprised the project's virtues. Citing the findings of three geologists, Koehler stated that in 43,700 acres to be worked in the immediate future, coal deposits were sufficient "to permit of an output of 3,500,000 tons a year for 100 years; with an enormous reserve tonnage in the balance of the tract—461,000 acres." Koehler predicted that "within a reasonable time," the enterprise could anticipate net earnings of 40 percent on annual receipts in excess of $850,000.[12]

Fisk and Robinson proceeded with a sale of securities. Frank Springer invested $90,000 and received $100,000 in bonds and $15,000 in preferred stock. With stocks and bonds in the hands of serious investors and cash in the hands of company officials, the enterprise proceeded from the realm of meeting rooms and paper to construction and operation. Surveys had already been completed and much of the needed right-of-way secured. A final element in the company's plans involved the purchase of the great Maxwell coalfield. The Maxwell Company received payment in the amount of $1.75 million for its most valuable asset.

In Koehler's letter of May 10, 1905, Frank Springer was listed as vice president and Charles Springer as treasurer of the St. Louis, Rocky Mountain and Pacific Company. When the new company's plans were revealed to the public in a front-page article in the *Las Vegas Daily Optic* in July 1905, neither was named among the company's directors, but the article identified Frank Springer as a trustee for the Maxwell Land Grant Company and one of its leading negotiators.[13] If his dual role in the transaction raised a question of conflict of interest in his mind or anyone else's, it apparently did not become an issue.

Having concluded the largest transaction in the history of the Maxwell Land Grant Company, and having recently endured criticism over the use of funds derived from the sale of holdings in Colorado, Springer traveled to the Netherlands to apply the proceeds of the coal sale to the company's obligations. Interest on the outstanding prior lien bonds was paid up, and the company was able to retire 601 of the 6,000 bonds authorized in connection with the issuance of the trust deed in 1888.

With its most valuable resources in the hands of other companies having the means and the will to develop them, the Maxwell Land Grant Company was a greatly reduced enterprise. In 1906, the trustees were advised to anticipate annual expenses of $22,500 and revenue amounting to approximately $27,000. The lull was only temporary, however, as new ventures, including a dam and reservoir and a railroad providing access to marketable timber in the Ponil country, were in the formative stages.

Change of Life

By 1905 Springer had reached that long-awaited time of life when he could withdraw from daily business obligations and devote more of his time to science. In that year Springer asked Kenneth Chapman, former art instructor at the Normal, to join him in Burlington for a period of two months, during which they would devote themselves to Springer's scientific work. In Burlington Chapman could observe Springer at the work he loved: "He arose about 11 A.M., looked over his morning's mail at his simple breakfast, and did not appear on the scene until noon. By that time the open air domestic stir was over, and there was little to distract him from his afternoon's work. But his real concentration came late at night when, long after the neighbors' lights were out, it was his habit to work in seclusion until 2 A.M. or after. With such an uninterrupted program, it seemed we accomplished more in a week than we could have done in months of occasional cooperation in Las Vegas."[14] At the end of the scientific interlude, Springer returned to New Mexico and his usual routine of business and legal work. His retirement from these activities was hastened, however, when he fell ill with a severe attack of heart disease.

The illness posed a grave threat, but it was also the subject of one of Springer's proudest achievements. Chapman retained a clear recollection of Springer's response to the crisis: "He called me to his house, and there, in his family's absence and with the marks of his illness still upon him, he told me of his plight, of his splendid rally after a night of despair when he faced about and determined to die game. 'We are going back to Burlington,' he announced, 'and we are going to fight this out to the finish.'"[15]

Springer was stricken in the spring of 1906. Of the season that followed, Chapman later remarked, "The memory of that summer in Burlington will never leave me, for if I had never realized his gameness before, I saw it then." Chapman accompanied Springer to Iowa, where he was both companion and training partner, as well as illustrator. Springer turned to the task of mending his health with the same powers of intellect and determination that he had given to the practice of law. He researched the matter, settled on a course of action, and gave himself wholly to the project. For once in his adult life, Springer was able to lay aside business matters and concentrate on an entirely different sort of task. From interviews with Springer, Ralph Emerson Twitchell described the therapeutic program:

> A study of the principal medical treatises bearing upon his ailment convinced him that a cure must depend chiefly upon the patient himself, not by medicines, but by plenty of well oxygenated blood, to be derived from carefully regulated exercise. Taking his artist, Mr. Chapman, he left New Mexico, "burning his bridges behind him," as he said, and went to Burlington, Iowa, where his scientific collections were, in which he could find the mental resource which Agassiz had predicted. There, in addition to a regular course of mild physical exercises adopted from the Swedish movement cure, he took to walking as his main reliance. Besides others which he did as the humor moved him, he had one course of a mile from his residence to the Mississippi River and back up an

> ascent of one hundred feet, on which he walked every day at a certain hour, rain or shine, with a definite object, which was to make the round trip without having to stop on account of his heart weakness. It took him just seven months to accomplish this.[16]

Perhaps unaware of Springer's physical condition, Gov. Herbert J. Hagerman offered an appointment that would have returned Springer to the board of regents of the normal university. From Burlington, Springer declined, explaining, "The condition of my health forbids my taking on any fresh labor or responsibility, and I am chiefly concerned with getting rid of what I have. I am paying the penalty of overwork, and it is imperatively necessary that I take a complete rest and respite from business, at a lower altitude, for a considerable time—a year or two, perhaps more."[17]

Springer and Edgar Lee Hewett had remained friends following their shared experience with the normal university. Upon completion of his studies at the University of Geneva, Hewett had accepted a fellowship with the Archaeological Institute of America and had engaged in archeological work in the Southwest and in Mexico. From Santa Fe he resumed regular correspondence with his loyal patron.

By December, Springer could boast of working until two o'clock in the morning without tiring. Having emerged from the prospect of a permanent and stifling disability, he reflected philosophically, telling Hewett, "I have the upper hand now, and believe I can keep it until some *new* breach is made in my defenses. It is really an interesting case from a medical and scientific standpoint. I saw my doctor in Chicago two weeks ago, and he told me frankly that what I had done cannot be done by medicine."[18]

As his heart improved, Springer increased the daily walk to a distance of ten miles, which he and Chapman could cover in approximately two and a half hours. In addition, Springer bought a parcel of land containing a fossil-bearing ledge near the river. Here he and Chapman worked part of each day during the summer, quarrying the limestone and extracting specimens from the layers of rock. "After about four years of this kind of work," Twitchell reported, "Mr. Springer found his heart as good as ever."[19] The heart may not have been "as good as ever," for it gave Springer trouble off and on over the remainder of his life. Springer found great satisfaction in his recovery and the lease on life it afforded, but he recognized that a watershed had been crossed, and that he would not be returning to the frenzied life he had known.

At work in Burlington, Springer was as relaxed as he had been in some time. His children were grown or nearly so, and his marriage, if not a model of romantic attachment, was stable. His investments were finally paying off and his professional ambitions had been realized. For the time being, he was content to roam the hills in search of fossils and good health, and to closet himself in the familiar environs of the brick museum. This quiet, introspective life, interrupted by occasional travels to New Mexico or one of the eastern cities, came as a welcome interlude following a long period of subordination to the demands of law and business. Springer's scientific studies remained the focus of his life in retirement from legal practice, but he soon developed new interests that reunited him with Hewett and enabled him to contribute to the development of a vibrant cultural life in his adopted territory.

CHAPTER 11

Dreamers and Doers

In an address delivered at the dedication of the new fine art museum in Santa Fe in 1917, Frank Springer spoke of the forces set in motion in the convergence of inspired vision and determined action. As he spoke, Springer was standing beneath a ceiling supported by massive wooden vigas, looking out over an auditorium fashioned in the style of New Mexico's mission churches. He had been the dynamic force in the effort to build the new structure, and had commissioned murals evoking the legacy of Saint Francis and the missionaries who bore his name. Said Springer, "It is when the doers and the dreamers work hand in hand; or better still, when the dreamer is himself a doer, that epoch-making events are achieved."[1] The throng assembled in the Saint Francis Auditorium was celebrating, as Springer was recalling, a full decade of effort devoted to the preservation of New Mexico's heritage and the stimulation of scientific and artistic endeavor.

In speaking of the dreamers and doers who had given life to New Mexico's emerging cultural institutions, Springer was building up to a rousing tribute to Edgar Lee Hewett, with whom he had now collaborated in a variety of pursuits over a period of more than twenty years. But Springer could as well have been the one characterized by these words, as doubtless many of those present knew.

Springer had enjoyed his association with Hewett through their mutual efforts on behalf of the New Mexico Normal University, and although that affair did not end as either might have wished, Hewett retained Springer's regard as a creative and capable administrator. After 1906 Springer continued as president of the Maxwell Land Grant Company, but he was officially retired from the active practice of law. Dividing his time between Las Vegas,

New Mexico, and Burlington, Iowa, he turned his attention to lifelong interests in education, science, art, and cultural affairs. If Springer devoted much time and energy between 1873 and 1906 to the building up of personal wealth and reputation, he spent much of the next two decades giving of his wealth and reputation to advance the causes in which he believed. A shared interest in the development of New Mexico's scientific and cultural resources brought Springer and Hewett together as trustee and administrator for a second time.

To the partnership Hewett brought creativity, promotional verve, administrative skill, and boundless energy. Springer was a possibility thinker who also had the personal means, collaborative skills, political connections, business sense, ingenuity, and competence to bring great plans to fruition. The relationship eventually extended over thirty years and resulted in the development of an array of vital educational and cultural institutions. Hewett carried on for some twenty years after the death of his senior patron, but in the view of Hewett's biographer, Springer's death in 1927 left "a void that could not be filled."[2]

A School for American Archaeology

Following the death of his first wife, Cora, in 1905, Edgar Lee Hewett threw his energy into the completion of his doctoral thesis. He also devoted time and effort to the struggle to preserve archaeological sites on the publicly owned western lands. These activities involved travel between East and West and visits to ancient puebloan sites in the Southwest, to Mexico, and back again to Washington, D.C.

Hewett had for some years been active in affairs of the Archaeological Institute of America (AIA). Through the institute, he made an influential friend in Alice Fletcher, perhaps the preeminent female American anthropologist of the nineteenth century.[3] While many of the institute's most eminent and entrenched members disdained American studies, she became an influential force for increased research in the West. With Fletcher's support, Hewett was chosen for the institute's fellowship in American archaeology and spent most of 1906 surveying cultural sites and meeting with local archaeological societies. In the course of his fellowship activities, Hewett visited Fletcher in Mexico and the two became allies. Fired with enthusiasm, Fletcher presented a plan of work to the American committee in

December 1906. The plan was accepted, and Edgar Lee Hewett was appointed director of American Archaeology.

In 1907 Fletcher succeeded to the chair of the American committee of the AIA. Persuaded of the need for a base for the institute's work in the West, she proposed at the December annual meeting that a permanent School of American Archaeology be created. The proposal was not greeted with unanimous enthusiasm, but it was approved. Hewett continued as administrator of programs in American archaeology, including the newly created school. A managing committee was named, with Fletcher as chair. Frank Springer was not one of the original members, but was added to the committee at a later date.

Several western cities expressed interest in hosting the new school, but by November, 1908, Fletcher was ready to name Santa Fe as the location, and she called a meeting of the American committee for this purpose. Santa Fe was the smallest among the cities vying for the school, but Santa Fe's unique character and proximity to significant cultural sites worked in its favor and the old Palace of the Governors was being offered as a home for the school. Moreover, Fletcher and Hewett were determined to have it there.

Fletcher faced opposition from a disgruntled eastern faction of the AIA. An initial vote went against continued development of the school, but Fletcher polled absentee committee members by mail, reversed the decision, and had Santa Fe declared the site of the new institution.[4] Fletcher thus triumphed for herself, for Hewett, and for Santa Fe, but resentments growing out of the dispute created deep and lasting divisions.

Hewett, in particular, did not go down well with Franz Boas of Columbia, Charles Bowditch, a supporter of the Peabody Museum at Harvard, and Alfred Tozzer, a Harvard anthropology professor. Defeated in the first instance, they nursed their resentments, and when the opportunity arose, they made their views known.

The Summer Field School

The American School was not a traditional educational institution offering courses and degrees. It was, rather, a school of research. Its students—mainly graduate students from leading academic institutions—learned practical methods of archaeology and interpretation of cultural materials through participation in work carried out under auspices of the school.

Field lecture, El Rito de los Frijoles. Edgar L. Hewett is addressing the group. Frank Springer is on the front row, third from right. Courtesy of National Anthropological Archives, Smithsonian Institution, Neg. 2003-39097.

A summer archaeology field school included laboratory activities in Santa Fe, lectures, and excavation at a nearby site.

The 1909 annual meeting of the managing committee was held in conjunction with planned work in excavating the Puye Ruins. Alice Fletcher was present and spent several days observing the work in progress. She also became better acquainted with Hewett's local supporters, including Frank Springer and John McFie. Soon after the meeting, Fletcher solicited the views of members of the managing committee concerning the field school. Springer responded with a statement in which he outlined plans for the school's development and offered observations on the field school operation. Coming to the likely object of Fletcher's request, he wrote as follows: "Answering categorically your question as to the character and efficiency of the staff employed in connection with the work as we saw it, I can say without hesitation that it represents the best type of scientific efficiency, and that for the means at the disposal of the Director I do not see how better results could be obtained."[5] In an apparent attempt to deflect criticism and rally support for Hewett and the school, Fletcher circulated the statement to AIA colleagues.

The field school, combined with a meeting of one or both of Hewett's boards, became a traditional event for the School of American Archaeology. Springer looked forward to the yearly retreat and especially enjoyed

Jordan Lummis, Frank Springer, Charles Lummis at El Rito de los Frijoles, 1911. Springer is leading Lummis, who was suffering a temporary loss of sight. Courtesy of Autry National Center/Southwest Museum, Neg. N-36541.

the companionship of Charles Lummis. In 1911 Lummis was able to attend the summer meeting, but he had suffered an unexplained loss of eyesight. Springer served as his guide, treading the trails of Frijoles Canyon with Lummis's hand on his shoulder. Looking to the following summer, Springer wrote in June 1912, "If your eyes should be right again—which may the good Lord graciously grant—we will blindfold you and take one trip, hand and shoulder, over the trail—just to show that it could have been done."[6]

Springer was partial to the Frijoles Canyon site, which afforded ample opportunity for exercise and fellowship as well as study. When Hewett determined that the summer school should investigate other sites, Springer lobbied to keep the encampment at the Rito, but in 1914 the work was concentrated on the Quarai site near Mountainair, New Mexico, and in 1915, the school was held at San Diego in conjunction with the Panama-California Exposition.

Springer found it possible to keep up the traditional encampment by serving as the leader of an expedition that surveyed pictographs along canyon walls of the Rito de los Frijoles in 1915 and 1916. In 1915 Kenneth Chapman, Carlos Vierra, Myrtle Boyle, Kate Mueller, Eleanor Johnson, and Santiago Naranjo, governor of the Santa Clara Pueblo, accompanied Springer to the Rito. Following up on earlier explorations by Chapman, the group went armed with artificial lights and ladders, examined some two hundred cliff dwellings, and recorded the types and numbers of drawings found there. The party copied more than a hundred drawings, including kachinas, human figures, masks, hands, animals, birds, and geometric figures.[7]

Left to right: Frank Springer, Santiago Naranjo, Kenneth M. Chapman, and Carlos Vierra at Rito de los Frijoles, 1915. Courtesy of Museum of New Mexico, Neg. 28087.

In 1916 a group including most of the same individuals returned to the Rito, this time taking flashlight photographs of many of the drawings and making plaster casts of figures that had been worked into the canyon walls. In early September the *New Mexican* reported that the Springer expedition had completed its work, having explored more than four hundred dwellings and having reproduced a total of 196 of the pictographs found there.[8]

Springer remained a member of the managing committee for the rest of his life, gave generously to facilitate the school's development, and took a personal interest in its accomplishments and political fortunes. Free of the demands of a high stakes legal practice, he could relax and enjoy the pleasant summer reunions, friendships with Lummis, Hewett, McFie, and others, and the satisfaction of seeing an institution blossom before his eyes.

In the busy decade between 1909 and 1919, Hewett collected an array of administrative titles, and he seldom gave one up to make room for another. Of necessity, his administration of the school led to a second title. In conveying occupancy of the old palace to the school, the territorial legislature decreed that the building should also house the new Museum of New Mexico. The museum was to be administered by the School of American Archaeology in accordance with policies adopted by a museum board of regents.

Two institutions, both administered by Hewett, made for pleasant company. If the arrangement complicated matters, it also afforded certain conveniences. But the legislature also specified that the Historical Society of New Mexico should continue to occupy rooms in the palace, and ere long it became apparent that three organizations, including one not directly under Hewett's exacting management, made a crowd.

The Old Palace

When Gen. Stephen Watts Kearny's troops put the Mexican governor to flight and took possession of New Mexico for the United States in 1846, the Governors' Palace on the north side of the Santa Fe plaza was already centuries old. A long, low structure facing the plaza, the adobe palace had long been the seat of government. The origin of the structure could be traced to approximately 1610, when the Spanish government was relocated from the upriver settlement of San Gabriel.[9]

When W. W. H. Davis arrived to serve as U.S. attorney in 1853, he found the old palace a sadly dilapidated seat of government. Observed the new official, "The stranger will be struck with the primitive appearance of these ruins: the roof is supported by a layer of great pine beams, blackened and stained by age; the floors are earthen, and the wood-work is heavy and rough, and in the style of two centuries ago." Upon his arrival in Santa Fe following a tortuous overland journey in 1878, Lew Wallace described the governor's office as "a picture of neglect and indifference." A few months later, Susan Wallace, could record her own impressions of "the leaky old Palacio del Gobernador." Despite the building's appearance, she saw it as the embodiment of a history "full of pathos and stirring incident."[10] Notwithstanding the first impressions of gentlepersons from Pennsylvania and Indiana, there were influential people in the territory who could appreciate the significance of the old palace. Springer's interest was motivated in part by personal experience; during his first term as a member of the territorial council, the council had met in its original chamber in the palace.

As governmental use of the palace waned, other organizations vied for use of space in the old capitol. The legislature granted occupancy of rooms in the east end of the palace to the Historical Society for offices, exhibits, and a library. From time to time other groups, including the Santa Fe Chamber of Commerce, were housed in the old palace. When Hewett returned to the Southwest following studies in Geneva, he became interested in the palace in relation to plans for the new School of American Archaeology. Discussions concerning such use of the palace began as early as 1907.[11] As an inducement for the Archaeological Institute of America to locate the school in Santa Fe, the legislature adopted "An Act to Establish a Museum for the Territory of New Mexico, and for Other Purposes."[12]

Former governor L. Bradford Prince grumbled that the proposed arrangement threatened the Historical Society's use of the building, but the *New Mexican* called the new museum "a magnificent idea" and waxed effusive about Hewett's plans. Following a presentation to the local Archaeological Society, the paper reported, "The Old Palace will be fitted up and decorated so as to attract attention the world over. Not with iconoclastic hand but reverently and always with a view of illustrating the period in which it was built and the period of civilization the particular room shall illustrate."[13] The museum would exhibit natural history specimens, works of art, historical objects, and models of prehistoric ruins.

The legislation created a board of regents to govern the museum. Three members were to be appointed from the managing committee of the School of American Archaeology. Alice Fletcher named R. W. Corwin, Charles

Museum of New Mexico regents at Puye Ruins, 1909. Left to right: John R. McFie, Frank Springer, Edgar L. Hewett, Charles F. Lummis, R. W. Corwin, and Nathan Jaffa. Courtesy of Museum of New Mexico, Neg. 13328.

Lummis, and Frank Springer to represent the committee on the museum board, which met for the first time on April 5, 1909. Other members of the new board were Judge John R. McFie, Territorial Secretary Nathan Jaffa, and Gov. George Curry.

Hewett moved quickly to comply with the legislative mandate and make the museum a reality. Jesse Nusbaum was appointed to superintend the renovation of the palace. Not until 1913 would the initial makeover be complete, but Nusbaum started in immediately to remodel interior spaces to accommodate Hewett's ideas for the museum. In the school's annual report for 1909, Hewett noted that alterations would be required in exhibit spaces, and that electric lights and cement floors would be added throughout the building. Looking to later stages of the project, he added, "As soon as funds can be secured for the purpose, the modern porch which finishes the entire front, facing the central plaza of the city, and which is the dominant architectural feature of the Palace, will be removed and replaced by an historic portal designed to conform to the architectural style of the period when the building was constructed."[14]

Hewett also arranged for Paris-trained painter Carl Lotave to travel from Denver and begin work on interpretive murals for the museum. Hewett's intention was to install the scenes in two central rooms, one depicting the Puye village as it might have looked when occupied, the other evoking the prehistoric site in Frijoles Canyon. Frank Springer sponsored the creation of the murals, giving a thousand dollars for the commission to Lotave.

By February 1910 Hewett was ready to show off the newly completed Puye Room. The *New Mexican* pronounced the museum "an attraction the like of which is not found outside of the very largest cities in the world." The paper acknowledged Springer's generosity and public spirit, adding, "The satisfaction to him must be great, aside from the fact that New Mexico will never forget the donor, nor the artist, nor the man who conceived the Puyé room of the cliff dwellers." A gala opening in August 1910, drew many of the territory's most distinguished citizens, along with other visitors. "Nowhere perhaps in New Mexico, or in the southwest this side of Los Angeles, could such a gathering be found," observed the *New Mexican,* "for nowhere is there such a lodestone to attract an assemblage of artists, scientists, political powers and literary men, and women, as that grand Old Palace."[15] The event was judged a success, as was the museum.

Hewett's roles as director of the School of American Archaeology and director of the Museum of New Mexico tended to blur, as projects, staff assignments, and budgets converged. His relationship with associates was sometimes strained by a tendency to take on additional responsibilities without relinquishing existing duties to make time for them. In 1911 Hewett was appointed director of exhibits for the Panama-California Exposition. This event, commemorating the opening of the Panama Canal, was to take place in San Diego. The exposition would be open throughout calendar year 1915, but there were extensive preparations to be made—obligations that would keep the director away from his regular duties much of the time leading up to and lasting through the event.

Despite opposition from Frederick Hodge and Alice Fletcher, both of whom had done much to advance his career, Hewett forged ahead, relying on the support of his unflagging admirers, Charles Lummis and Frank Springer. Hewett's involvement in San Diego did not end with the exposition, but led to his appointment as director of a new museum in the city. In accepting the position, Hewett had no intention of giving up those he already held in Santa Fe.

Statehood

Due in large measure to the increasing proportion of time Springer was spending out of the territory, his involvement in the final push for statehood was minimal. He had taken a hand in an earlier unsuccessful attempt, helping draft a proposed state constitution in 1889, but he did not take part in the constitutional convention of 1910, as did his brother Charles.

Upon being furnished with copies of the document to be placed before the voters, Frank Springer offered his impressions for readers of the *Las Vegas Optic.* He gave the proposed constitution his unqualified endorsement and congratulated those who had crafted it, observing, nonetheless, that such a document is by nature, "a body of compromises of conflicting opinions. . . . Of course," Springer observed, "there are persons who would find fault with the Sermon on the Mount, if delivered by one of the opposite political faith, and it is not surprising to hear discordant war cries from some quarters about this constitution—the chief one being that 'the convention was controlled by the corporations.'" Having anticipated one of the main objections to the document, Springer proceeded with a rebuttal, observing, "If it was, then I must say that the finished product is the greatest example of self-denial in all history. For it is evident to any one who can read plain and straightforward language, and will do so honestly, that instead of retaining any advantages for themselves, they delivered themselves, hand and foot, and for all time, to the control of the people's elected agents."[16]

The new constitution was approved by a margin of more than two to one, but the desire of some citizens for more direct access to the machinery of decision-making led to a proposal to amend the document. The proposed remedy was the "Blue Ballot Amendment," mandating a more flexible amendment process. The amendment provided for popular consideration of an unlimited number of constitutional amendments on a blue-tinted ballot separate from that used to vote on candidates for public office. Congress had already granted the petition for statehood, but with a proviso requiring that the new state provide for consideration of a less restrictive amendment process in its first general election; the struggle over the Blue Ballot Amendment had the effect of prolonging the constitutional debate.

Speaking at a Republican rally in Las Vegas on October 7, 1911, Springer entered the debate, calling on citizens to reject the proffered amendment. He urged on listeners the notion that one of the purposes of a constitution is to provide for a stable form of government. In his view, the process proposed by advocates of the Blue Ballot Amendment would have the effect of reducing fundamental provisions for government to the level of ordinary statutes. "If this be the best way," he said, "then the conclusion inevitably follows that there is no need of a constitution at all; and that is the practical result which every thoughtful citizen, in the last analysis of this question, must face." Springer made a similar speech at Raton and later acknowledged authorship of an opinion piece that appeared front and center in the November 4 edition of the *Las Vegas Optic,* titled "What a Practical Business Man Thinks about the Blue Ballot."[17]

Despite the efforts of Springer and other opponents, New Mexico voters approved the Blue Ballot Amendment by a substantial majority. Opposition to the proposal may have been the undoing of the Republican Party in the general election of 1911, for the party lost key races for governor and one of the two seats in the U.S. House of Representatives. The dissension evident during the campaign was largely dissipated in the euphoria that prevailed when, on January 6, 1912, New Mexico shed decades of frustration and joined the Union as the forty-seventh state.

Tempest in Santa Fe

Having accepted the position of director of exhibits for the Panama-California Exposition, Hewett began spending time at San Diego and elsewhere in preparation for the event. In Kenneth Chapman and Paul A. F. Walter, he had capable staff members who could manage operations in Santa Fe. Nonetheless, Hewett wanted to remain firmly in control of all enterprises within his purview. In August 1913, he returned to Santa Fe from an extended sojourn and walked into one of the more contentious episodes of his career.[18]

Having recognized their city as a place of cultural and historical interest, the businessmen of Santa Fe undertook to promote tourism through a newly energized Chamber of Commerce. The chamber had organized itself into a number of functional committees, including a "New–Old Santa Fe Committee," the purpose of which was to encourage adoption of a characteristic style of architecture suggested by Pueblo and Spanish colonial structures.

On the evening of September 16, 1913, the chamber met in the old palace, where offices of the organization had been located for about two years. Bronson Cutting was present and reported for the publicity committee. Ralph Twitchell and former governor Prince were also on hand and were consulted as authorities on historical matters. A suggestion that Santa Fe proclaim itself "the oldest city in the United States" and use the slogan on promotional materials met with general enthusiasm and received a perfunctory nod from the resident historians. It was proposed that envelopes bearing the boast be printed, and before long, orders had reached 68,000, later increased to 125,000. The scheme—and the proclamation of Santa Fe's claim of unrivaled antiquity—headlined the *New Mexican*'s coverage of the meeting the following day.

The validity of the boast was highly doubtful, and there were those present at the meeting, other than Twitchell and Prince, who knew history and could not in good conscience remain silent for the sake of a relatively harmless excess of boosterism. Sylvanus Morley, an archaeologist on Hewett's staff, called the group's attention to the questionable nature of Santa Fe's claim but was politely ignored. Hewett also was present, but reportedly said nothing. Instead, he went to the chamber president, Harry Dorman, the following day to insist that the false claim not be printed on the chamber's envelopes. Further, he requested that the chamber pay back rent due for use of its office in the palace or vacate the premises. Hewett's actions enraged members of the business community and precipitated a chain of events that threatened his continued administration of the school and museum.

Weary of Hewett's badgering and possessed of a backlog of grievances, Dorman responded with a frontal attack. Bronson Cutting was pleased to be of assistance as only the owner of the local newspaper could. On October 3, the *New Mexican* ran an interview with Dorman about the chamber's plan to publicize the city to prospective tourists, but the article's most newsworthy feature was a broadside calling for Hewett's replacement.[19]

The *New Mexican* persisted in its attacks on Hewett, enlisting the aid of his academic rivals—Alfred Tozzer and Roland Dixon of Harvard, and Franz Boas of Columbia—who were only too happy to furnish critiques of Hewett's qualifications.[20] The paper went on to question the credibility of the school's managing committee. The *New Mexican* listed the committee's thirty-one members by name and occupation in order to make the point that many were not, as Lummis had claimed in the school's defense, "University Men," and that few members were equipped to judge Hewett's qualifications as a scientist.[21]

On October 30 the executive committee of the school rallied to Hewett's defense, meeting at the U.S. National Museum in Washington, where chairman W. H. Holmes was curator of anthropology and Springer was in residence as an associate in paleontology. The committee adopted a statement expressing confidence in Hewett and asserting that his activities were as contemplated in the original plan of operation for the school.[22]

A rumor not entirely discouraged by friends of Hewett had the School of American Archaeology moving from Santa Fe, possibly to Los Angeles. Dr. J. A. Rolls of the Archaeological Society in Santa Fe wired Springer to ask, "Do you believe recent attacks of the New Mexican could result in removal of the school?" Rolls indicated that local groups were prepared to adopt resolutions in support of Hewett if needed. Replied Springer, "Answering your wire, Santa Fe may depend upon her real friends to protect her interests in the matter. But Santa Fe and New Mexico owe to their own self[-]respect and their reputation with other parts of the country to publicly condemn these attacks on an eminent scientist, one of their own citizens, actuated solely by personal resentment of a success the authors were unable to achieve."[23]

Springer's statement was intended for public consumption. It was read to an assembly of the Archaeological Society of New Mexico, which then adopted resolutions in support of Hewett and the school. For the further edification of Rolls, Springer wrote a longer letter detailing the issues and relating what he believed to be the true and dishonorable motives among Hewett's academic detractors—principally, petty jealousies dating to Hewett's appointment as director of the institute's program in American archaeology.[24]

On November 10 Twitchell could report that the attack on Hewett seemed to be effectively squelched and that if it continued, a "mass indignation meeting" would be held in Santa Fe.[25] The momentum shifted in Hewett's favor when the Chamber of Commerce met in a contentious three-hour session on November 14, emerging with a resolution that read, in part: "Be it Further Resolved, That the Santa Fe Chamber of Commerce disclaim any connection with or responsibility for any criticism of the director of the School of American Archaeology at Santa Fe, heretofore made in the public press or otherwise."[26]

On December 18 the executive committee of the school again met in Washington. In a statement drafted by Springer and adopted by the committee, the group referenced the attacks on Hewett and expressed the committee's appreciation of supportive letters, telegrams, and resolutions received from organizations and individuals in Santa Fe. "They express the entire confidence in the management of the School as hitherto administered," the document read, "and show that the attacks referred to are vigorously resented." That said, the committee affirmed its support of Santa Fe as the location of the school and declared its satisfaction with the relationship between the people of New Mexico and the school.[27]

The affair was not quite ended, as Dorman, representing those who were still dissatisfied with Hewett, approached the Managing Board of the Archaeological Institute of America at its January, 1914, meeting in Montreal. Dorman reviewed the grievances concerning Hewett and urged a reorganization of the school, but his pleadings were ineffectual and the board concluded by affirming its support of the managing committee and Hewett.

Notwithstanding the successful defense of Hewett, there was a good deal of right on the side of his detractors. A case certainly could be made—and was made—that Hewett slighted his responsibility to the School of American Archaeology and used his position as director to sustain him in an array of diverse and somewhat peripheral activities. It can also be argued that the struggle over the school's direction simply amounted to a clash of divergent interests. While the program supported by Springer and Hewett may not have catered to the desires of the Ivy League anthropologists and their Santa Fe allies, it gave substantial impetus to the development of an extensive state museum and did much to promote the region's antiquities as cultural and economic assets.

CHAPTER 12

Springer of the Crinoids

The loss of his collaborator, coming at a time that found Frank Springer still immersed in law and business, occasioned a predictable decline in his scientific productivity. The loss of Wachsmuth did not, however, alter a career plan that had long since taken shape in Springer's mind, nor did it compromise his fundamental commitment to science.

Following Wachsmuth's death in 1896, Springer saw the massive three-volume monograph on the Crinoidea Camerata through to completion, but not until 1900 did he publish another substantial paper. A monograph published by the Museum of Comparative Zoology in 1901 proved that Springer could turn out brilliant work on his own. The *London Zoological Record* noted in a review, "The work of the year is Springer's magnificent Monograph on the Structure and Relations of *Uintacrinus*."[1]

The fossil collection and library assembled by Springer and Wachsmuth eventually outgrew the space available in Burlington. The volume of Springer's published work also grew, and with it his reputation among scientific colleagues, one of whom praised his work in print, calling him "Springer of the Crinoids."[2]

The Collection

After Wachsmuth's death Springer took on a few smaller projects, added to the fossil crinoid collection, and began a transition from law and business to the life of a serious scientist. In Wachsmuth's absence, he found others with whom he could exchange information and ideas, and on whom he could rely for sound critical advice. He cultivated scientific correspondents at home and abroad, but with the exception of one brief collaborative effort, he worked alone.

One of the few who could fully appreciate Springer's work was Charles Schuchert, a leading academic geologist. Like Springer and Wachsmuth, Schuchert was essentially self-trained. He served an apprenticeship as an illustrator and assistant under several prominent geologists and in 1893 was employed in the U.S. Geological Survey. In 1904 Schuchert was invited to join the faculty at Yale University.

Springer also maintained cordial relations with Alexander Agassiz, and his publication of *Uintacrinus* under Agassiz's auspices reflected a continuing emotional commitment to Harvard and the Museum of Comparative Zoology. For some years Springer nurtured the idea of placing the fossil collection at Harvard and moving to Cambridge to pursue his own research. He persisted in this notion despite reservations about the quality of care Agassiz could offer for a large collection.

As early as 1890 Springer had observed an alarming lack of care given the items in Agassiz's museum. Some exhibits were without identifying labels, and some items previously on display seemingly had disappeared. Reporting his observations to Wachsmuth, Springer wrote, "The trouble is Agassiz has amassed collections disproportional to his ability to provide competent care for them, and they have in many cases been left—I suspect—to the tender mercies of janitors and carpenters."[3] Despite this discouragement, Springer remained hopeful that circumstances would improve and that the museum might yet make a desirable scientific home for his collection and himself.

The matter was on Springer's mind when he wrote to Agassiz in February 1902, to lay out his ideas. Springer saw the potential for Harvard to build "a great Echinoderm Collection, which by reason of the extent and character of its material, and facilities for study, should be *the* Echinoderm Collection of the world, and would make the Museum the chief repository of facts in that branch."[4]

Springer had purchased Bernardina Wachsmuth's interest in the Burlington collection and had invested an estimated $12,850 in fossil material. He expressed willingness to donate the entire collection to the Museum of Comparative Zoology. He was frank to share with Agassiz not only his vision but also his concerns: "There is no use doing it, however, simply to pile up racks-full of specimens and stuff the already congested Museum further. With no other prospect than that, I should not feel much interest. My idea is not only to bring together these great Crinoid collections of my own and what the Museum already has, with such others as can be bought, but also your Echinoid Material, and all recent Echinoderm collections made under direction of yourself and your father, to constitute a special and distinctive feature of the M.C.Z. To carry out this idea," Springer continued, "the first desideratum is a place to house and utilize such a collection. There is not room enough in the Museum now even for what is already there; and, what is a still greater objection, it is not a fire-proof building." Springer suggested the construction of an additional wing to the museum to provide the needed exhibition, library, work and study spaces. Looking to his retirement from law, Springer expressed his intent to assist in cataloging the collection "without desiring any compensation whatever."[5] It may have seemed to Agassiz that Springer was asking for the moon. Had Agassiz fully appreciated Springer's capacity for turning large dreams into realities, he might have parlayed Springer's interest into a major acquisition for the museum. Instead, the idea languished.

By 1906 little progress had been made at Cambridge and Springer had begun to consider more promising alternatives. On his return from a trip to Europe in November 1905, Springer arranged to spend a day with Charles Schuchert at the Peabody Museum at Yale. In the course of a day's pleasant shoptalk, Springer broached the subject of his quandary concerning an appropriate disposition of the collection. According to Schuchert, Springer's major concern was proper care and housing for the material—a need which, in Schuchert's view, was beyond the means of any university. Knowing of plans for a new building for the U.S. National Museum in Washington—the Smithsonian's repository for natural history materials—Schuchert suggested this institution as a possible base for Springer's scientific work.[6]

Springer may have given passing consideration to the Walker Museum at the University of Chicago and the American Museum of Natural History in New York, but having given up on Harvard and the MCZ, his attention was soon focused on the U.S. National Museum.[7] Springer's intended course of action was strongly intimated when, in 1906, he purchased the W. F. Pate collection, consisting of more than fifty thousand specimens, and presented it to the U.S. National Museum.

The U.S. National Museum

In 1904 the Smithsonian Institution broke ground for a new Natural History Building across the mall from the original Smithsonian castle. The project took several years to complete, but the result was an edifice worthy of a nation's pride. Exterior walls of red brick were faced with blocks of

U.S. National Museum, Natural History Building, Washington, D.C. Courtesy of Smithsonian Institution Archives, Neg. 26560.

granite, ten inches or more in thickness. The building provided a flexible space that could be allocated to exhibits, laboratories, offices, and storage of collections as needed. Moreover, the structure was designed to be fireproof. The building first opened to the public in 1910, but it was not fully occupied for several years. In 1957 the U.S. National Museum was formally divided into a Museum of History and Technology and a Museum of Natural History, and in 1969, the natural history museum assumed its present name, the National Museum of Natural History.[8]

As Springer turned his attention to the U.S. National Museum, matters converged to make an alliance with the museum an attractive proposition. Springer was well connected at the Smithsonian. Charles Doolittle Walcott, secretary of the Smithsonian, was a geologist and had long been aware of Springer's work. Walcott was in a position to offer the resources Springer required. Ray S. Bassler, assistant curator in the Department of Geology, also understood the value of Springer's work. He was responsible for working out the details of Springer's relationship with the museum.

Springer was interested in an apparent convergence of his work and that of Austin H. Clark, a young Harvard graduate. While Springer concentrated on a wide-ranging morphology of the fossil crinoids, Clark was pursuing similar research on extant species. Springer believed that the National Museum could benefit from a coordinated effort encompassing both areas. He was sufficiently interested in this possibility that he offered to pay a portion of Clark's salary if Clark could be induced to join the museum staff. In April 1907, Walcott wrote to advise Springer that he had talked with Clark and found him receptive to the notion.[9] Clark was brought on as an honorary collaborator in the Division of Marine Invertebrates in 1908 and given the more substantial title of assistant curator in 1909. When a Division of Echinoderms was created in 1920, Clark became its curator, remaining in the position until his retirement in 1950.

By 1908 Springer had settled on the National Museum as the repository for his collections and the base for his research activities. In the new museum building and the amenities offered in support of his work, Springer found all that he had sought. Although the building was not yet complete and the bulk of his collection would remain in Burlington for some time, Springer formalized his commitment to the museum and received assurances concerning plans for housing the collection and his research. Ray Bassler wrote to apprise him of accommodations to be provided: "In our arrangement of space in the new building, Dr. Merrill [George P. Merrill, curator of geology] and I have assigned two good sized rooms to you overlooking the park and the Capitol, for laboratory and general work rooms, exclusive of the regular exhibition space which we discussed before. This will give a room to both you and Clark, and will also provide you with a place for your library. Besides this there is plenty of well lighted space in the basement where we expect to store collections and do all the heavier preparatory work." Also at Springer's disposal were an office and a photographic facility. "I hope you will not consider yourself an outsider," Bassler wrote, "especially since I am always thinking and planning with you in mind as an insider, and I also hope that you and Clark will have many years to enjoy the echinoderms."[10]

Sunset on the Potomac

Early in 1911 plans were made for moving Springer's collection to Washington. Bassler was detailed to oversee the transfer. Prior to his departure for Burlington to supervise the packing and shipping of the specimens, he wrote to inform Springer of an organizational change that was of interest to both men. The message also reflected an effort on Springer's part to ensure favorable treatment for the study of invertebrate fossils in the organizational structure of the museum. Said Bassler: "Some days ago I said that we would hear of other changes in the Museum [of Natural History] at Burlington together. However, Mr. Walcott thought I better have mine before I go away, so that, starting today, a department of Paleontology, with yours truly as Curator in charge, and with the previous departments of Vertebrate Paleontology and Paleobotany reduced to sections, is created. Now just see what you went and done with your letter. That I appreciate your efforts is unnecessary to say, but more of this later." [11]

The bulkier portions of Springer's collection were transported to Washington by chartered express car in July and the remainder was shipped in November. The *Burlington Hawk-Eye* noted the occasion with an article titled "Farewell to Crinoids," characterizing the collection as "the largest and most complete in this specialty to be found in any public or private museum in the world." [12]

In December, Secretary Walcott expressed his pleasure at the consummation of the long-anticipated move. "Now that you have selected Washington as your scientific headquarters and will become more closely identified with the activities of the National Museum," he wrote, "I greatly desire to see your name actually connected with that establishment. It would, therefore, give me much pleasure if you would accept the honorary title of Associate in Paleontology, which I hereby tender you in recognition of your important contributions to this branch of science, and in consideration of the many valuable donations you have made to the national collections." [13] By year's end Springer had mailed a printed postcard to correspondents, advising them of his change in location.

As a permanent residence for Josephine and himself, Springer took a suite at the Dresden, an elegant apartment house built in 1909. Located in the fashionable Kalorama Heights, the Dresden was one of Washington's desirable addresses, and was just over two miles from Springer's laboratory in the Natural History Building. [14]

Once installed in his new home, Springer spent the greater part of his time in Washington, with extended sojourns to New Mexico on business of the Maxwell Land Grant Company and the school and museum in Santa Fe. Kenneth Chapman visited Washington in 1912 and again in 1914 and found Springer comfortably situated and up to his old habits: "Provided with every convenience, he had quite adjusted himself to his new surroundings, except for one circumstance. The inflexible rules of the National Museum obliged

The Dresden apartment building in Washington, D.C. Courtesy of Photographic Archives, National Gallery of Art, Goode-Phillips Collection, Neg. GP 6069.

every inmate to vacate promptly at 5:30 P.M., just as he was laying his plans for a good night's work. However, with his family residence established in Washington, he was able to carry on his night work as of old, in the comfort of his own home."[15]

In Washington Springer carried on an active program of research and publication. He contributed brief notes, articles, and reviews to scientific journals, but most of his work was focused on a few major projects. He continued to sponsor fieldwork each summer, sending collectors to gather fossil material in support of his research objectives.

Springer had for some years been building toward the publication of a classification of the Crinoidea Flexibilia. He had begun this work in 1896 as the next logical step following publication of the work on the Crinoidea Camerata. When he began his study of this group of fossils, few examples could be found in the world's museums, but his research revealed an abundance of species—some 176 that were described in the monograph as published.[16] As an example illuminating Springer's character, Chapman later recalled an incident in connection with his work on the Flexibilia monograph. When Springer traveled to New Mexico for the annual retreat at the Rito de los Frijoles in 1917, he brought along a large portion of the text, intending to give it a final reading for corrections. "But he was absorbed with other matters and returned to Santa Fe without having given it his attention," Chapman recalled. "Through some preoccupation he mislaid it."[17]

A thorough search for the envelope containing the missing manuscript was conducted, but to no avail. Springer returned to Washington facing the unenviable task of reconstructing some one hundred pages of text. The missing material was integral to a study to which he had devoted much of the previous fifteen years. By the following summer the manuscript was again in shape for the printer.

It was, then, with considerable embarrassment that Chapman discovered upon unpacking Springer's camp outfit—which he had packed for storage in the basement of the fine arts museum—that the missing envelope was tucked away in Springer's camping gear. "I have known remorse," Chapman recalled, "but never such as I suffered for the next two days until I could tell him of my part in the lamentable affair. I begged him to make as vigorous comment on the subject as the situation seemed to demand, and I knew well his capability in that direction. Instead, he took it calmly, and, after what seemed an age of reflection, said simply, 'Well, it's of no use to us now; I have spent the best part of the last year re-writing it.' And then, opening the envelope and finding something else within it, he added, 'By George! at least we have saved some good sheet music.'" Having perused the found manuscript, Springer later called on his guilt-ridden assistant to thank him, assuring Chapman that the new work was so far superior to the old that "it would have been the greatest possible mistake to have published it in its original form."

Springer's *Monograph of the Crinoidea Flexibilia,* a two-volume work consisting of 486 pages of text with numerous figures and plates, appeared in 1920 under the imprint of the Smithsonian Institution. It was reviewed favorably in the *Geological Magazine* of London, the *American Journal of Science,* and other scientific periodicals.

In 1920 Frank and Josephine disposed of unneeded furnishings and closed out the family home in Las Vegas. As indicated on the title pages of subsequent scientific publications, Springer remained loyal to his adopted city and was proud to claim East Las Vegas as his official residence for the remaining years of his life.

Dam Builders

With the creation of the St. Louis, Rocky Mountain and Pacific Company and the transfer of the Maxwell Company's coal interests to the new corporation, the activities of the Maxwell Land Grant Company were significantly reduced. The expansion in coal production created an increased demand for mine timbers, however, and stimulated investment in timber operations. With Springer and other investors, Thomas Schomburg organized the Continental Tie and Lumber Company and the Cimarron and Northwestern Railway Company for the purpose of harvesting timber in the extensive drainage of Ponil Creek, north of Cimarron. By early 1908 twenty-two miles of track had been laid, rolling stock had been purchased, and the companies commenced operating in North Ponil Canyon. As stands of timber were cleared in the north, stretches of track were taken up and laid in the Middle and South Ponil Canyons. By 1930 depletion of timber and increased use of logging trucks had eroded the line's business and the road was abandoned.

Land sales continued to provide some income, but the gold reserves in Baldy Mountain were regarded as the Maxwell Land Grant Company's best hope for increased revenue. In 1909 the company committed to new explorations on Aztec Ridge. Results were inconsistent and mostly disappointing. The mines produced more than $350,000 in gold in 1915, but depletion of high grade ores and increased labor costs combined to end the brief period of prosperity.

In contrast to the marathon meetings of Springer's early years as president of the Maxwell trustees, meetings in the early decades of the twentieth century were brief, sometimes bordering on perfunctory. The board had been purged of eastern influence and Springer had consolidated control of the company in men he trusted. Upon the death of Francisco Manzanares in 1905, banker Daniel Hoskins was elected to take his place. J. M. Cunningham and Oliver Lyman Houghton, both of Las Vegas, also served as trustees with Springer and Jan van Houten. This group remained intact from 1905 until 1922, when Cunningham died and Houghton resigned in ill health.

The board's major object during these years was the retirement of the prior lien bonds, representing obligations to investors who had exchanged the old Maxwell Company's securities for new bonds when the company was reorganized in 1888. Payment of these obligations required that the company generate a substantial surplus, and excess revenues were not easy to come by. Springer rarely missed a board meeting and was routinely returned to the office of president.

In 1906 Charles Springer set in motion an ambitious project involving the construction of a dam at the head of Cimarron Canyon. Above the canyon lay the broad Moreno Valley, an expansive *vega* situated at an elevation of approximately nine thousand feet. The valley provided good grass for cattle and other livestock, and supported some farming during a brief growing season. Cutting through the Cimarron range for eons, the Cimarron River had captured the vast watershed of the valley. At the head of Cimarron Canyon was a rocky passage at a place called Eagle Roost Rock.

The potential for impounding a large volume of water behind a dam at Eagle Roost Rock had been recognized at least since 1888, when Levi Preston surveyed the area and suggested such a possibility.[18] In 1893 the Maxwell trustees had discussed possible construction of a dam and reservoir, along with ditches and improvements sufficient to irrigate some forty-five thousand acres. The trustees agreed that the project was not one for the company, but they were amenable to selling the dam site and potential farmland to investors willing to undertake such a venture.[19] Nothing came of the discussion, and the idea lay dormant for the next several years.

In more than twenty-five years of living and ranching on the plain east of the Cimarron Mountains, Charles Springer had seen for himself that significant acreage could be cultivated with sufficient water. He also knew of difficulties the Maxwell Company had with its ditch systems, dependent as they were on lesser streams and subject to extreme fluctuations as the seasons changed. By damming the Cimarron and creating a reservoir in the Moreno Valley, it would be possible to provide a larger and more reliable source of water for irrigation. Lands worth only a few dollars an acre for grazing would be worth much more if made suitable for agriculture. Charles had evidently given much thought to the matter, and in February 1906, he filed a notice of intent to appropriate water from the Cimarron. In April he approached the Maxwell Company trustees, presented his plans, and was awarded an option on approximately six hundred acres needed for the dam site.[20]

Frank Springer was by now in retreat from new business commitments, but he helped finance the venture. Charles Springer, with associates Hilton M. Letts and Frederic Whitney, organized the Cimarron Valley Land

Company in 1907, for the purpose of carrying out the Eagle Nest project. Of seven thousand shares of stock issued, one thousand were conveyed to the Charles Springer Cattle Company in payment for twenty-eight thousand acres of land. The remaining shares, other than nominal shares conveyed to the other directors, were divided equally between Frank and Charles Springer. Charles took immediate steps to get the project under way, but obstacles were encountered and a decade passed before construction could begin.

For one thing, the new company did not own or otherwise control all the land to be flooded by the dam. Some landowners in the valley could be induced to sell needed parcels, but three ranchers refused to give up their land. The Cimarron Valley Land Company brought suit to condemn the property in 1909, but a protracted legal proceeding ensued and the matter was not resolved until 1916, when the court ruled in Springer's favor and the families were forced to sell.

Construction of the dam was financed through a sale of bonds. The San Antonio firm of Bartlett and Ranney was engaged to design the structure and Neal Hansen, a native of Denmark, was employed as construction engineer. As the dam took shape, workers downstream prepared the

Eagle Nest Dam. Photo by Jesse L. Nusbaum, Courtesy of Western History Collection, Denver Public Library, Neg. N-65.

distribution system, building diversionary works near the confluence of the Cimarron River and Cimarroncito Creek and constructing canals, ditches, and laterals.[21]

Construction began in earnest in 1917. Labor was scarce and competition with nearby mines exacerbated the shortage. The war in Europe also diverted men and materials from domestic projects, but the dam was substantially complete by June 1918, and the lake began collecting water shortly thereafter.

Frank Springer spent several months in New Mexico that summer and fall of 1918 helping manage the CS Cattle Company, while Charles and Frank's son Hank gave much of their time to war work. He watched as the project took shape and by November 15, was able to offer Henry Porter a report on its progress:

> The Eagle's Nest Dam was finished according to plan, and has now about 12,000 acre feet of water, forming quite an impressive looking lake up into the valley where the remnants of the old Dugan house are just visible above the water. We were able to supply water to the Springer Ditch people, and the French colony, in season to save their crops, of which they have an abundance. The dam is a very fine structure, and I do not think is excelled anywhere for strength and absolute security, and for the efficient impounding of water. It is a monolithic structure of the best type, solid concrete arch, keyed into the granite walls on either side—with spillway through, and over, the solid rock. It is 117 feet from creek level to bottom of spillway, and 124 ft. to top of dam. Five feet of the spillway could be used, if necessary—as no harm would result if water flowed over the top. It has been approved by the State Engineer, and certified to have a capacity of about 85,000 acre feet. How much water a season's accumulation will produce is what we have to find out by experience, and we are not going to be in a hurry about selling land until we find out more about this. We have paid off one maturing installment of bonds—$35,000—and are prepared to take care of all the rest, without selling any land. So we can afford to bide our time, and await the progress of economic development following the war, which is sure to be of absorbing interest.[22]

The postwar prosperity did not materialize as anticipated, and Frank Springer was forced to advance large sums from his personal fortune in order to meet the company's obligations. The debt incurred in construction of the dam was not fully paid until 1942.

The Eagle Nest project was intended as a sound business investment that would return a profit to its owners based on the enhanced value of agricultural lands. Prior estimates of acreage to be watered ran from 45,000 acres to as many as 70,000 acres, but if the lake ever watered 30,000 acres, the area under irrigation certainly did not exceed that. By the mid-1920s, Charles was ready to concede that the project was a failure. In truth, its better days were yet to come. By the early 1920s, the village of Therma was attracting tourists and fishermen and supported a growing community of vacationers. The community was renamed Eagle Nest in 1935 and became a popular seasonal resort and vacation spot.

The shining lake in the Moreno Valley was Charles Springer's vision, and Charles supplied the push to bring about its realization. Involved as he was in various pursuits in other places, Frank Springer was able to be helpful in his capacity as president of the Maxwell Land Grant Company, but his main role in the project was that of an investor and partner. The venture was, however, indicative of his capacity to think big and long term, and of his willingness to undertake substantial risk in the spirit of free enterprise.

CHAPTER 13

The New Museum

Comfortably settled in Washington, with the National Museum as his scientific base and the Dresden as his family home, Frank Springer assumed a pattern of activities that found him working steadily through much of the year and vacationing in New Mexico two or three months each summer.

As had been the case throughout their married life, Springer and Josie remained firmly and affectionately attached, but going in different directions much of the time. During frequent times of separation, Frank and Josephine remained in correspondence, exchanging chatty letters, concluding with "Love and Kisses" and other expressions of endearment.

Whatever Josie's early expectations of her marriage, she adjusted to Springer's habits, making separate plans and enjoying the perquisites of affluence in her own way. She welcomed the companionship of her daughters, visiting Helen in Buffalo, Laura and Ada in Philadelphia. Eva became an accomplished painter of miniatures, studying at the Art Students League and in Paris, sometimes living with her parents in Las Vegas and later in Washington. In company with one or more of her daughters, Josephine traveled in Europe, spent time on the West Coast, and enjoyed the amenities of city life. She sometimes shopped for the men, sending socks, ties, handkerchiefs, and other necessities to her husband and sons. In his state of semiretirement, Springer could devote more attention to family, spending time with his sons in New Mexico and traveling to New York to welcome the return of his wife and daughters from a European holiday.[1] To their children, Springer and Josie were affectionately known as "Pater and Mater."

While in New Mexico Springer continued to divide his time between the ranch at Cimarron, his adopted hometown of Las Vegas, and Santa Fe, but he increasingly found himself drawn to Santa Fe. There he enjoyed the company of artists, scholars, and political and business associates. His presence also facilitated his efforts as the political mainstay of the cultural enterprises administered by Edgar Lee Hewett.

As Springer's time in the city increased, so did his civic involvements. Friends who had known him only by way of his notoriety as an attorney, businessman, and benefactor of worthy causes, became better acquainted with the whole man. At ease, Springer revealed himself as a person of considerably greater breadth and depth than his fellow citizens may have imagined.

A Many-Sided Man

In October 1916, a columnist for the *Albuquerque Morning Journal* recalled a memorable evening some years before on the Rito de los Frijoles. Charles Fletcher Lummis had offered a rendition of "Mandalay" and natives of San Ildefonso Pueblo had performed the Eagle Dance. "Then came a lull broken suddenly by the sweet strains of a flute playing on one of the crags of the opposite canyon wall, high above the caves and communal ruins. 'Schubert's Serenade' never sounded so entrancing as it did that night, played by a man who might justly be called 'New Mexico's Foremost Citizen,' for his name and work are familiar to scientists the world over, and who as a patron of art and education leads 'all the rest' in the Southwest."[2]

The writer went on to report that Frank Springer had recently given a recital at the home of Edgar L. Hewett. Accompanied by Mrs. C. E. Doll, Springer offered a program ranging "from selections of classic brilliancy to songs of the heart." The writer declared the evening "one of those rare occasions that become unforgettable memories in life," and, said the paper, "One got a glimpse of another side of a beautiful character." Though occupied with weighty matters throughout his career, Springer had never relinquished his avocation as a flutist, sometimes participating in community musical performances during his years in Las Vegas.

The *Journal* writer was further surprised to find Springer an authority on Oriental rugs. A question calling for such expertise had arisen, and Springer, it happened, "knew more about Bokharas, Daghestans, Sirez and other rugs than anyone else in the capital." Springer had acquired his knowledge as a collector of fine rugs. In a 1915 inventory of his assets, Springer reported that the rugs had cost him approximately twenty-five thousand dollars, all paid out of odd sums that could be spared for a private extravagance. "I have had more than enough pleasure in the mere possession of these masterpieces to fully justify the cost," Springer noted.[3] He considered the collection a realizable asset, but hoped that the rugs might eventually grace the homes of his grown children.

Springer was an outdoorsman as circumstances permitted, hunting fossils along the Mississippi in Iowa, fishing the streams of New Mexico and Colorado during his early years in the territory, and enjoying "tramps" around Burlington and in New Mexico into his seventies. He was enough at home on a horse to ride out a parcel of land added to the mountain ranch of the CS Cattle Company in 1918, and he took considerable pride in a photograph of himself, his grown sons, Charles Springer, and J. M. Cunningham atop Wheeler Peak with their mounts.

As a youth and college student, Springer enjoyed ice-skating; as a resident of Las Vegas, he could sometimes be seen skating with a younger crowd on the frozen reservoir in Gallinas Canyon. In retirement and throughout his infirmity, he was a baseball fan, following the progress of the major leagues, as Kenneth Chapman recalled, "from the first ball over the plate to the wind-up of the World's series, with all the interest that he gave to the copper market, the Philippine question, or the early struggles of aviation."[4]

Springer made his mark in science as a paleontologist, but an exotic wildflower that he collected near the Rito del los Frijoles and turned over to a botanical acquaintance entered the floral taxonomy as *Mentzelia springeri,* or Springer's blazing star.

Springer was frequently called upon to commemorate momentous occasions in the life of his community, territory, and state. He spoke at commencement exercises for the normal university and the University of New Mexico. When Theodore Roosevelt attended a reunion of Rough Riders in Las Vegas, Springer was asked to offer remarks and pin a medal on the visiting hero. He delivered memorial addresses honoring John Marshall, William McKinley, and Roosevelt. He spoke to commemorate the dedication of the state's new art museum and to mark the groundbreaking for a new auditorium at the normal university. Though his starched prose makes somewhat ponderous reading, his orations consistently drew praise from the local press.

Springer had a first-rate education, an unquenchable curiosity, and tremendous energy, bounded only by the limitations of the heart disease that encroached on his activities in later years. These assets served him well and allowed him to bring to widely varied situations a perspective that was comprehensive of natural and human history. He was as much interested in the ancient civilizations of Rome, Greece, Egypt, and Mesopotamia, as of the prehistoric cultures of New Mexico. Of his interest in the Mediterranean and the Near East, he remarked, "No part of my personal library is better filled than that relating to the literature, the history, and the explorations of that region; and the best thumbed book I possess is the Poems of Homer."[5]

There was, then, ample justification for the characterization of Hewett and others who occasionally referred to Springer as a "many-sided man."

Donald Beauregard and the Saint Francis Murals

As Frank Springer settled into a comfortable retirement, he became increasingly amenable to providing support for worthy projects proposed by Edgar Lee Hewett. He routinely subsidized museum salaries, sponsored research, provided funds for purchases of needed property, desirable works of art, and research materials, and could be counted on to support a special publication or promotional venture. From time to time Hewett enlisted Springer's interest in an artist in need of support.

In April, 1912, Hewett wrote to acquaint Springer with the prospects of twenty-seven-year-old Donald Beauregard, a native of Utah and a graduate of the University of Utah. Beauregard had been employed as supervisor of art for the city schools of Ogden, Utah, but had given up teaching to make a career as a painter. Beauregard had come to Hewett's attention during the summers of 1909 and 1910. A correspondence developed, and Beauregard disclosed his desire to study art in a serious way. Hewett concluded that Beauregard had a talent worth nurturing, and in time he approached Springer for assistance: "You will remember very well, I think, Mr. Donald Beauregard, who was with us at the Rito in 1910, and will recall his rather exceptional ability as an artist. Beauregard burned his bridges behind him last year, gave up teaching, and went to Europe determined to get in at least a year or two of work under proper instruction. He has kept me advised of his movements from time to time. I am enclosing a letter from him which I should be glad if you will read. I have very great confidence in Beauregard's future."[6] Hewett expressed hope that a way might be found to help Beauregard carry out his plan for working and studying in Europe, preparatory to taking up his career in the States.

Springer replied on May 20, expressing interest. "I feel much inclined to lend him a helping hand on the basis he suggests," Springer wrote, "*viz,* furnish him a year's expenses, in exchange for a year of his time later, which could be distributed in some way so as not to tie him down too much to his detriment." Springer authorized Hewett to proceed with arrangements for Beauregard's support. By early fall of 1912, the matter had been arranged and Springer wrote to say that he would see the young artist through a year of study abroad. Springer expressed his view concerning the expectations that might accompany his commitment. "It was suggested in former correspondence that in return for support now he might give some of his time after he returns," Springer stated. "I would prefer this, on some equitable basis, rather than to treat the transaction as a mere loan of the money. I shall be glad to devote the greater part of the resulting output to the extension of our scheme of mural paintings,—retaining a little of it for some things of personal interest which I should like to have done."[7]

Beauregard wrote from Spain to thank his new patron. "Your help comes at a critical moment," he wrote. "I was 'broke.'" Noting that he had been broke before, Beauregard explained that his impoverished condition likely would have meant the end of his hopes for a career in art. In consideration of Springer's interest, Beauregard told about his early life and the decision to pursue painting. A native of Fillmore, Utah, he had lived on a ranch until the age of sixteen, at which time he left home, eventually enrolling at the University of Utah.[8]

Beauregard traveled to Paris, where he took up formal studies in anatomy and decoration. By December he had to report that he was feeling "rather washed out," and that he had been advised by a physician to retreat to a higher altitude as a precaution against tuberculosis. He moved on to Switzerland, where he could continue to pursue his studies while enjoying the therapeutic benefits of the mountain climate. Beauregard later traveled to Munich, but he again fell ill. He advised Springer of his imminent return to the United States.[9]

Beauregard landed in New York and proceeded to Washington, where he met Springer and accepted a commission for a series of murals depicting scenes from the life of Saint Francis of Assisi. Hewett had already broached this possibility, and Beauregard had prepared sketches during his European travels. Following the meeting with Springer, Beauregard traveled to his parents' home in Utah, then to Santa Fe.

On December 28, 1913, the *Las Vegas Optic* announced that Springer had given ten thousand dollars to the School of American Archaeology "to immortalize on canvas the chief scenes in the life of New Mexico's patron saint, Saint Francis of Assisi." Springer's gesture was hailed as "one of the largest gifts ever presented to a New Mexico educational institution" and "the largest gift ever set aside for art in New Mexico." Hewett's original plan called for the murals to be displayed in the New Mexico Building at the Panama-California Exposition in San Diego, then installed in the Palace of the Governors. When it became evident that the murals would not be finished in time, it was decided that they would be incorporated in plans for the new art museum.

Beauregard made himself at home in space provided in the old palace and began by assembling frames, stretchers, and other materials.[10] By February 10, 1914, the project was under way and Beauregard had nearly finished the center panel for one of three triptych murals, *The Renunciation of Santa Clara*. It was the only piece completed by Beauregard, for by March 6, he was experiencing symptoms that could not be ignored. He wrote Springer, telling of trouble with his stomach and a growth on his neck, diagnosed by a local physician as a cystic goiter. Before receiving Springer's letter urging him to proceed to the Mayo Clinic, Beauregard left Santa Fe to consult doctors in Denver.[11]

He underwent surgery on March 14, and on March 27 wrote to convey the prognosis: "At last a definite decision has been reached about me. I am sorry to tell you it is not a happy one. I have waited to write until now in order to put it before you plainly. I told you [in a letter dated March 18] the Pathologist had taken the mass from the stomach for examination, and finally he sent the report that it is cancer. I can't tell you the scientific name, but it shows two kinds of cancerous growth both of which, says Dr. Perkins, are bad. Now that seems to upset about as much as anything can the plans that we have developed about my work, etc." Springer was moved by the artist's plight, telling Kenneth Chapman, "It does surely seem as if the gloom were getting thick enough to cut with a knife, but we must whet up our knife and cut. Except for the death of my early frontier comrade in Mexico thirty years ago, and the desperate illness of Laura in Rome, nothing has affected me so deeply as Beauregard's brave letter in which he announced his own death sentence."[12]

When he was well enough to travel, Beauregard went home to Utah. He had written Chapman in Santa Fe concerning the shipment of his personal belongings. Paintings, sketches, and other work were to be held at the museum for Springer.[13] Beauregard wrote from Fillmore to inform Springer of his arrival, expressing a good deal more cheerfulness than his condition warranted. He had been home just over two weeks when he died on May 2, 1914.

Kenneth Chapman and Carlos Vierra were chosen to finish the murals, making use of concepts developed by Beauregard. Despite the complications in bringing the work to completion, the finished murals made an impressive feature in the new auditorium, drawing attention not only to the life of Saint Francis but also to the zeal of the Franciscan missionaries. In the view of one observer, "This exercise in historical pedigree wrapped the Fine Arts Museum in an aura of religion and European culture and conferred moral and intellectual authority on the institution, including Hewett and Springer."[14]

A Place for Art

Once established in the old Governors' Palace, the Museum of New Mexico began accumulating treasures and exhibiting them, supporting the research and educational activities of the School of American Archaeology, sponsoring artists, and otherwise expanding to fill the space available. Before long, Hewett became persuaded of the need for a building to accommodate a permanent collection of art and support a growing community of artists.

It is to Hewett's credit that he recognized the importance of Santa Fe's emergence as a center for art and artists. From the beginning, he and Springer nurtured this development through the museum's practice of buying and exhibiting the works of local artists. Hewett's involvement in the Panama-California Exposition further stimulated the growth of art in Santa Fe as he commissioned works evocative of Southwest cultures for the New Mexico exhibits.

Hewett had worked with architect I. H. Rapp to design an exhibit building evoking the characteristic mission architecture of the pueblos; the same concept would be used in the design and construction of the new art museum in Santa Fe. Along with fellow regent John McFie, Springer plied the legislature for support. By March 30, 1915, he could tell Jesse Nusbaum, "I put in a couple of months at Santa Fe getting the bill passed, and a still bigger job is ahead of me in raising the outside money and securing the

Meeting at the Palace of the Governors concerning construction of the Museum of Fine Arts. Kenneth Chapman (standing) and Edgar L. Hewett (seated) are at the far left of the photo. Seated at the table to the right of the elevation drawing are (l–r) John R. McFie, Frank Springer, William C. McDonald, Napoleon B. Laughlin, and Paul A. F. Walter. Photo by Jesse L. Nusbaum, Courtesy of Museum of New Mexico, Neg. 14235.

site."[15] The legislation authorized funding in the amount of thirty thousand dollars, to be matched with an equal sum in private contributions.

Springer took charge of the fund-raising effort and approximately one year later, he made good on the commitment of matching funds. A ceremony was staged at the Palace of the Governors on April 12, 1916, in conjunction with the meeting of the Museum of New Mexico regents. In the presence of the regents, state and local officials, and friends of the museum, Springer presented a check in the amount of thirty thousand dollars. Hewett, Springer, and James Seligman were appointed to constitute a building committee.

The funds raised by Springer were presented in the name of the School of American Archaeology, probably in order to bolster the school's standing with local citizens and policy makers. It has been widely assumed that all or most of the money came from Frank Springer's own funds, but his notes indicate a personal contribution of five thousand dollars to the building fund, with another twenty-five thousand in "contributions of personal friends secured by me."[16] Donors were named in annual reports of the school and museum, and in published accounts in the local press.

The building site was obtained through an arrangement with the Santa Fe school board, and the Old Barracks Building, a relic of the abandoned Fort Marcy, was razed to make way for the museum. Following work with Hewett preparing exhibits in San Diego and a brief period of employment with the Fred Harvey Company, Nusbaum returned to Santa Fe to superintend the construction. Additional land was needed to ensure adequate space for construction, so Springer purchased the adjoining lot and home, conveying title to the portion needed for the museum. The home, left from Fort Marcy's "officers' row," was remodeled for use as the director's residence.[17] Springer held a mortgage for some years before canceling it and making a gift of the property.

The building's design borrowed heavily from the adobe missions at nearby pueblos. Brick and cement were the main construction materials, but the facade evoked the Pueblo mission style and recognizable elements of specific churches. The general floor plan was also suggested by the layout of one or more of the old mission churches. The museum would include an auditorium seating more than seven hundred people, exhibit galleries, offices, and a reception hall.

In October 1916, Hewett reported that construction was proceeding and completion was expected by June 1917. Hewett was still heavily involved in San Diego, where the Panama-California Exposition had been extended through 1916. Paul A. F. Walter, Kenneth Chapman, and Nusbaum remained in Santa Fe, minding the museum and overseeing construction of the new edifice. In December, Walter reported, "Progress on the new building has not been as rapid as I had counted upon, but what has been done looms up more majestically and impressively than had been dreamed of. The entire community seems to be more and more struck with the beauty of it all."[18]

Over a period of approximately eighteen months, Santa Fe residents watched as the building site at the intersection of Grant and Palace Avenue was transformed. The old military barracks came down and a basement was excavated. Foundations were laid, and walls of brick and mortar rose to impressive heights. Finally great timbers were laid in place atop the walls of the auditorium. Vigas, *latias,* and corbels were put in place, doorways and windows fashioned to imitate comparable features of the Pueblo missions. Plastering, landscaping, and interior finish work brought the structure near completion.

The building was finished in early fall and dedication ceremonies were held on Sunday afternoon, November 25, 1917. The event marked the emergence of a regional identity reflecting the sensibilities of relative newcomers but incorporating the landscapes and native cultures of New Mexico. As the principal speaker, Springer acknowledged the conflicting emotions stirred by the war overseas and expressed his hope that the building might serve as an inspiration, drawing attention to a wealth of indigenous materials for art. "Depending no longer upon other lands or times for inspiration to brush, to chisel, to trowel or to song," he proclaimed, "we shall find at home the themes for boundless achievement, and our arts shall grow—as this temple has grown, and as all true and enduring Art must ever grow—straight from our own soil."[19]

The euphoria surrounding the opening of the new Museum of Fine Arts carried over into the new year, 1918. Springer and friends seized the momentum generated by the occasion and pushed ahead in pursuit of a still more ambitious vision for the development of Santa Fe.

The Santa Fe Style

Visitors to Santa Fe may be surprised to learn that the city as surveyed from the downtown plaza is not the result of careful preservation of historic structures and a congenital preference for adobe. Rather the romantic Santa Fe ambiance is the result of efforts to cultivate an appearance of antiquity related to the region's history and prehistory. The story of Santa Fe's deliberate transformation is well told by Chris Wilson in *The Myth of Santa Fe: Creating a Modern Regional Tradition.* Frank Springer, though at most a part-time resident of Santa Fe, played a major role in efforts to establish the Spanish-Pueblo or Pueblo revival style as the dominant architecture in Santa Fe.

The movement to adopt a consistent architectural style was motivated at least in part by the desire to enliven a struggling economy. With the "Plan of 1912," leading boosters promoted a vision for the city as an attractive tourist destination with a distinctive appearance. The plan called for the consistent application of a motif incorporating adobe construction, *portales,* exposed vigas, *canales,* and other features of Pueblo and Spanish architecture.

Proponents advocated the construction of public buildings modeling the "New–Old Santa Fe" style.

The establishment of a Pueblo-inspired style of architecture required undoing several decades of taste-making that were driven by opposing values. Merchants who had come to Santa Fe from the eastern United States and Europe aspired to demonstrate a progressive spirit in ways that would be recognizable to the people back home. With the coming of the railroad, they began making Santa Fe over in earnest, applying imported notions of civilization and progress. By the turn of the century, the Italianate, Queen Anne, and Second Empire styles of architecture had made their appearance, adding to a local repertory that already included Greek, Romanesque, and Gothic revival.[20] Brick and stone were favored as building materials of choice.

Hewett and Carlos Vierra became enthusiastic promoters in public utterances and in print, urging Santa Feans to embrace styles rooted in regional tradition. In 1917 the School of American Archaeology published "Our Native Architecture in Its Relation to Santa Fe," by Vierra, and "Santa Fe in 1926," by Hewett. The papers reflected a tendency on Hewett's part to favor promotion over scholarly rigor in programs of the school, but they offered provocative arguments in support of the movement. Hewett laid out a blueprint for the transformation of downtown Santa Fe to reflect the proposed style and declared, "A determined community can bring the whole thing to pass by 1926."[21]

Frank Springer was an early and ardent supporter, beginning with his involvement in the restoration of the Palace of the Governors and his leadership as chief backer of the Museum of Fine Arts. He was among the financial guarantors of *Art and Archaeology,* an elegant magazine published by the Archaeological Institute of America, and he took great pleasure in an issue devoted to Santa Fe and the new museum in January–February, 1918. Of special interest was an article by Carlos Vierra, proclaiming the advent of a distinctive American style based on the dwellings and mission churches of the pueblos. Springer found it remarkable that Santa Fe should receive such positive exposure by virtue of the distribution of the magazine to a nationwide audience of ten thousand subscribers.[22]

From his post in Washington, Springer was well positioned to influence plans for a new federal building in Santa Fe. Although it was understood that construction could not begin before the end of the war in Europe, federal officials were at work in the spring of 1918, considering possible locations and preparing architectural plans. The plans were of much interest to Springer, who advised Paul Walter in February 1918, of a recent visit with L. A. Simon of the Supervising Architect's Office in the Treasury Department. Springer felt certain that he had won a friend in Simon, who had given assurances concerning his support for use of the local style. "He is thoroughly in sympathy with our ideas as to architecture in Santa Fe—for hotels and all public buildings &c," Springer wrote, adding, "I think he will be the one man whose views will control."[23] Springer requested and received photographs illustrating the proposed style.

Springer stayed on the case and reported further progress in April. Simon had reviewed the photographs sent at Springer's request and had been won over to the proposed style. "He was intensely interested, as this was just what he wanted," Springer reported. "He is thoroughly imbued with the possibilities of doing a big thing out there on entirely new lines for government work, and has *no idea of anything else,* and he assures me that there is no probability of interference with his designs." In Springer's mind, the building would constitute "another monument to a definite epoch in the development of the Southwestern part of the country.[24]

Springer's hope for an unqualified triumph was thwarted by the nature of the site selected for the building—a half city block across from the Saint Francis Cathedral that did not permit a setback from the street or accommodate the notion of a decorative rear facade. Nonetheless, Simon and associates produced a design incorporating the mission motif and elements of the Governors' Palace. Construction ensued and the building was completed in 1920.

Springer contributed to the cause in a more personal way when he entered into an arrangement with Carlos Vierra involving construction of a residence modeling the New–Old Santa Fe style on land owned by Springer. Springer requested that the museum support the effort by allowing Vierra the loan of its cement mixer, wheelbarrows, and other tools. If surplus building materials could be made available, Springer would pay for them at cost. Arrangements were made and Vierra began work, functioning as architect, contractor, and superintendent. The finished structure was significant not only for its design, which featured a modern interior camouflaged by an exterior reflecting the Pueblo style, but because it carried the movement out of the historic plaza area and onto the high, dry mesa to the south.[25]

The effort to maintain the Santa Fe style has continued for several decades, reinforced through government-sanctioned designation of historic properties and a detailed code of regulations. Springer did not seek or desire recognition for his role in the movement to secure general acceptance of the New–Old Santa Fe motif, but he did as much as anyone to popularize the architectural style and civic values that came to characterize modern Santa Fe.

Reorganization of the School

In 1916 the managing committee of the School of American Archaeology found it advisable to incorporate, rather than continue as an operating unit of the Archaeological Institute of America. Frank Springer was asked to investigate the legal requirements for incorporation. He conveyed his findings to William Henry Holmes, chairman of the school's managing committee, in letters of December 4 and December 6, 1916. Springer noted that incorporation was a legal necessity if the school was to acquire and hold property for the purpose of generating an endowment. Along with the change in legal status, the managing committee proposed a change of name, from School of American Archaeology to School of American Research.

Springer attempted to anticipate and answer possible objections to the proposed changes. To the charge that the school was compromising its commitment to archeology, Springer replied that one could hardly study archeology without involving related sciences and that the school had been doing so already with the full knowledge and approval of the institute. A second objection was that the school was attempting to separate itself from the AIA. Although there was patently some truth to this, Springer denied any intent on the part of the school to pull away and pointed to matters in which the institute retained authority.

In December 1916, the managing committee forwarded the proposed changes to the AIA Council and the council gave its approval on December 30. Springer prepared the required legal documents and the school filed for incorporation in New Mexico. By February, 1917, the School of American Research had been established as a corporate entity and the former managing committee became the school's new board of managers.

By 1918 much had been accomplished by Hewett and his loyal patron. Hewett had survived challenges to his leadership and was now more firmly in control of his multifaceted domain. Near the end of 1918, however, the enterprise nearly lost its leader when Hewett considered a possible appointment as director of the U.S. National Museum. Aware that the Smithsonian Institution was in the market for a director of the museum, Springer had brought Hewett's name to the attention of Secretary Walcott. Springer was prepared to help Hewett obtain the position, and the two exchanged correspondence on the matter for several weeks. Hewett met with Secretary Walcott in San Diego, and again in Washington.

On November 11 Springer passed on as much intelligence as he could. According to his understanding, Walcott's interest in Hewett was in relation to the directorship of the U.S. National Museum, meaning all Smithsonian museums, excluding arts and industries. "This would be a most admirable arrangement," Springer advised, "and I think we should definitely decide that if this is what it means you will take it if he wants you to."[26]

From all indications the appointment was Hewett's for the taking, but he declined. Hewett held other special appointments from time to time, but he never relinquished the dual roles that were at the heart of his work as promoter, builder, administrator, scientist, and educator. With the completion of the Museum of Fine Arts, the tandem of Springer and Hewett had perhaps realized its greatest achievement, but the two would continue for some years more, plying their varied and complementary talents in the interest of the school and museum, the City Different, and the cause of creative endeavor in New Mexico and the Southwest.

CHAPTER 14

Profit and Loss

Following the opening of the new Museum of Fine Arts, Frank Springer enjoyed another half dozen years of relatively good health in which he could pursue his crinoid studies, continue his support of the cultural enterprises to which he was devoted, and tend to essential business of the Maxwell Land Grant Company. As president of the Maxwell trustees, he presided over a much-diminished enterprise, the major activity of which seemed to be selling off pieces of itself in order to pay off the ever-present prior lien bonds held by the Dutch investors.

At this time in his life, Springer did not need to work, either for money or for reputation. But he was energized by the work he enjoyed and he had goals yet to attain, so in spite of increasing physical limitations, he worked. He published two articles in 1918, a brief research note in 1919, and the great two-volume monograph on the Crinoidea Flexibilia in 1920. Springer appeared in the scientific literature at least once in each of the following years, publishing substantial papers in 1921 and 1923. He also began another major study, on the Silurian crinoids of North America.

During these years, it was Springer's pleasure to reap a harvest of honors and goodwill, the consequence of sustained contributions on many fronts. But he also suffered gracefully a number of losses, many of which are normal concomitants of advanced age—the loss of friends and family members, and of robust health.

In appreciation for their support of working artists, the Taos Society of Artists conferred honorary membership on Frank Springer and E. L. Hewett in August 1918.[1] This put the two men in exclusive company. Among those designated as active, associate, and honorary members, there were only twenty-one in all over the society's entire twelve-year history; Springer and Hewett were the only people designated as honorary members.

The Great War

On July 28, 1914, Springer wrote his research colleague Austin Clark from New Mexico, where he was spending the summer. "It looks as if the clash for which Europe has been so long preparing was at hand," Springer observed. The archduke of Austria-Hungary had been assassinated by a dissident Serbian nationalist, and war clouds had been gathering ever since. "The crucial test of armies and navies, and the remodeling of the map of Europe, may be upon us with startling swiftness," Springer wrote. "But it is a dreadful thing to contemplate."[2] Three days later, Germany declared war on Russia, and the fight was on.

At his advanced age, Springer could do little for the cause, but he could contribute to the war work, endure philosophically the peril of his sons who served at the front, and endeavor to lift the spirits of his countrymen and correspondents. In the wake of the American declaration of war, Springer wrote to his British colleague F. A. Bather, sending proofs and plates of a forthcoming publication. In addition to the usual shoptalk about crinoids, he commented on the war and expressed doubt that his fellow citizens knew what they were in for: "History is being made so fast nowadays that it almost takes our breath away trying to keep up with the big events that crowd one another from one day to another. That the English-speaking peoples are in arms together is certainly one of the biggest of them. That they have a tremendous task ahead is fully realized by thoughtful men—but not at all by the general public and the politicians. A few Zeppelin raids would do good here as they did in England."[3]

By the time Springer delivered the dedicatory address for the new art museum in November 1917, the war had been raging for more than three years and the United States had joined the conflict, declaring war on Germany on April 6. Springer did not apologize for the peaceful celebration in which those assembled were engaged, but declared that it was important to affirm mankind's higher impulses even as the barbarism of war threatened their existence. "The world," he said, "is facing the question whether modern civilization is to survive or perish; whether it shall endure the trials which beset it now, and remain the dominant influence in human affairs, or whether it shall go down before the fiery blast in which the forces of man and nature are harnessed for purposes of destruction, leaving only a mighty ruin to mark the most stupendous tragedy of all time." While brave men fought, Springer contended, it was the duty of those at home to keep the lights of human intelligence and creativity burning brightly.[4]

The Springers became more personally involved when their oldest and youngest sons, Edward and Wallace, were commissioned as lieutenants in the U.S. Army. At the urging of his family, the middle son, Hank, accepted the exemption provided for managers of enterprises that were vital to the war effort. He continued as general manager of the CS Cattle Company but served on the local Council of Defense and as a member of the draft board.

Edward was placed in command of a field artillery battery and Wallace was trained as a bomber pilot. By early summer both had been sent overseas, and Springer was anxiously awaiting news from the front, telling Bather, "I have two sons with the American forces in France, both Lieutenants, one in the Air Service and one in the Field Artillery, and it is very slow work hearing from them, notwithstanding it is said that military mail has the preference."[5]

Along with friends across the country, the Springers began an uneasy surveillance of casualty lists appearing in the local paper. It was perhaps inevitable that they, or friends close to them, would suffer the loss of a son. Springer steeled himself for this eventuality, adopting "the stoic's philosophy" and a determination to confront great loss with equal courage.

On July 11, 1918, Springer wrote his family a long letter from the ranch in New Mexico. He told of a trip to Las Vegas with Hank on July 3 and of the "Cowboys' Exhibition" they had taken in on the Fourth. Having exhausted all the pleasantries that came to mind, he came to the somber purpose of his letter. "I have left until the last the most unpleasant news," he wrote. "Last evening Dr. Cunningham called me on the telephone to say that word had been received that Leonard Hoskins had been killed in action. No other particulars."[6]

Leonard Hoskins was the son of Daniel T. Hoskins, a business associate in Las Vegas, and a fellow trustee of the Maxwell Company. Springer wrote the elder Hoskins immediately to extend his condolences and share the philosophy with which he and Josephine faced the possibility of a similar loss. "Mrs. Springer and I have considered this thing, and talked it over calmly, and have agreed that for us, and for our friends in like situation, these are times different from any that we could have imagined in the happy years when we were bringing up our families, and that they call for different emotions and sentiments from those with which we were then familiar. And we decided that whatever might happen, we would bear our losses with equanimity—not making them the occasion of futile mourning, but rather taking comfort in the fact that we have brought up healthy and high-minded boys, to form a part of the dependable manhood of the country whenever its time of need should come, and that in dying gloriously for it they have simply repaid, in the highest manner that man can pay, what the country has done for them."[7]

Springer took the occasion of Leonard Hoskins's death to counsel courage on the part of his own family, telling Josie and the daughters, "We must be prepared to face what comes, and meet the inevitable, if it should come to us, with the same courage and fortitude that our boys would wish us to show, and without emblems or evidence of mourning, except to reverently affix a gold star, where it belongs."[8]

The war held no such dreaded consequence for Frank and Josephine, but they did receive with alarm the news that Wallace had been hit by enemy fire while taking part in the Allied offensive on Saint Mihiel and the Argonne. Not until December 11, 1918—a full month after the armistice—did the Springers learn of the matter, via a telegram from Wallace. Wallace minimized the injury, but he was hospitalized for a month and then sent to a convalescent camp. Upon his discharge by the military doctors, Wallace returned to the United States and was mustered out of the service. Edward was at first assigned to the forces of occupation, but he too was soon discharged and on his way back to New Mexico.

Having survived the war with their family intact, Frank and Josephine were shocked and grieved when Hank died suddenly on January 30, 1920. The irony of Hank's death was not lost on Springer, who wrote to Bather in April, telling of the painful loss and the consequent interruption of the crinoid work. "It is a curious example of the decrees of Providence," he wrote, "in that whereas my other two sons who went to the war and were in the thick of it at the front returned with only minor wounds, this one, who was elected to remain at home because he was the indispensable manager of a food-producing industry, falls in the prime of health and vigor before a swift attack of Pneumonia following Influenza."[9]

In a memorial brochure sent to friends, Springer told of the son who, unlike his brothers, did not earn a degree or train for a profession. Instead, he followed a vocation that seemed to choose him: "He was from the

Springer sons at CS Ranch headquarters, c. 1919. Left to right: Lew Wallace, Henry (Hank) Strong, Edward Thomas. Courtesy of CS Cattle Company.

beginning, and of his own deliberate choice, a potential ranchman—not merely as a means of future livelihood, but because he loved animals, and loved the broad fields and open skies."[10] Hank had studied agriculture at the University of Illinois. With his parents' blessing, he left school and went to work on the ranch, eventually taking its management over from his Uncle Charles. At the time of his death, Hank had traveled far in his aspiration to build a ranching operation modeled on scientific principles, and he had won the respect of employees, neighbors, and cattlemen, as his father said, from Chicago to Mexico.

The loss was especially painful to Springer, who confided to Jesse Nusbaum that he had never been able to get close to Wallace as he did to his other boys, "especially Hank, with whom I was as intimate as fathers ever get with their sons."[11]

Reconciliation

The events of 1903, in which Edgar Lee Hewett and Frank Springer parted company with the New Mexico Normal University, were not particularly acrimonious. Hewett moved on, landed on his feet, and fashioned a stellar career in Santa Fe. Frank Springer quietly withdrew as an active participant in affairs of the Normal, but remained a vocal and visible citizen-supporter.

In 1919 Springer had the opportunity to effect a more overt reconciliation with the school in which he had invested so much hope and devotion. He made a point of taking Hewett along on a sentimental journey to the scene of earlier triumph and disappointment. The occasion was Springer's participation on behalf of Charles Ilfeld in plans for the construction of an auditorium on the campus.

From a hopeful beginning, the university had grown to an enrollment of nearly fifteen hundred students. Facilities that had been at first spacious now were strained to accommodate the school's needs. Among the priorities was an assembly hall in which the full student body could meet. Appropriations totaling seventy-five thousand dollars were sought and received from the state of New Mexico, and in October 1917, the regents authorized Pres. Frank H. H. Roberts to proceed with plans for the new building.[12] In April 1919, the board engaged an architect, E. W. Hart, who later asked that the Santa Fe firm of Rapp, Rapp and Hendrickson be associated with him in the project. Plans were drawn according to the needs of the campus and a shortfall in funding was evident. Former regent Charles Ilfeld became involved at this point, with Frank Springer serving as his representative.

October 9, 1919, was, in the view of Hugh W. Price, "a momentous day in the history of the Normal University."[13] The day began with a meeting of the regents at which Springer announced that Ilfeld would donate the sum of twenty-five thousand dollars needed to complete the building project.[14] The regents approved resolutions specifying the board's obligation relative to the gift and providing that the new building should be called the "Ilfeld Auditorium," in memory of the donor's late wife.

A ceremonial groundbreaking took place in the afternoon, and in the evening the community celebrated with a banquet at the Castaneda Hotel. Charles Ilfeld was guest of honor and Springer served as master of ceremonies. Attending were Gov. Octaviano Larrazolo, Hewett, President Roberts, the regents, and other dignitaries and citizens of Las Vegas. Springer addressed the guests, reviewing the early history of the institution and developing the context of the project at hand. He professed gratitude for opportunities afforded his children to attend the Normal and declared, "I look back on that period with a sense of deep obligation, and if I had anything to do with the movement which furthered the creation or operation of this University, I feel it is one of the greatest privileges of my life to have done so; and for anything I have contributed I have been repaid a hundred times."

Following remarks by Governor Larrazolo and several community leaders, Springer spoke in honor of Charles Ilfeld and announced the gift tendered earlier in the day. This drew "an indescribable outburst of enthusiasm" on the part of the crowd, as well as grateful responses from O. O. Blood, mayor of Las Vegas, and President Roberts. Finally the governor offered a motion: "That in recognition of Mr. Frank Springer's untiring efforts toward the up-building and success of the Normal University, extending over nearly a quarter of a century, it be declared the sense of the meeting that the present Normal building should hereafter be known as 'Springer Hall.'" The motion found unanimous support among those present and was ratified by the regents on December 1, 1919.

Plans were approved in the spring of 1920. Bids were solicited, and a contract was issued for construction of the auditorium, an administration building, a dining hall, and a heating plant. The auditorium was placed in service in an unfinished condition in 1922, but the project was plagued with unanticipated costs. A full decade passed before the building was completed, with the help of additional state appropriations and gifts from the Ilfeld family.

Laurels

In 1918 Springer reached an age that, among friends, he referred to as the "scriptural time limit"—that is, three score and ten. At seventy, Springer was in relatively good health and nowhere near ready to take his ease. Over the next several years, he was, however, the recipient of honors bestowed in recognition of sustained contributions in varied fields. He also took on a new responsibility, succeeding William Henry Holmes as chairman of the board of managers of the School of American Research in 1922.

Springer was the recipient of two honorary doctoral degrees. The first, a doctor of science degree from George Washington University, was conferred in 1921. The second, the doctor of philosophy, was conferred by the Friedrich-Wilhelms University in 1924. Springer delighted in sending Eva a photographic copy of the diploma, along with his own English-language transcription of the German diploma. He added, "I am really much pleased to receive such a degree from that University, which is one of the greatest in Germany,—as it was entirely unsolicited, and I especially like the last reason given."[15] The citation concluded by noting that Springer's success in science had been attained outside his primary vocation in life.

In 1922 Springer was honored for his sustained generosity to the School of American Research and the Museum of New Mexico. Friends had commissioned sculptor Cartaino Scarpitta to create a bust to be presented to the State of New Mexico in recognition of his contributions. On the evening of September 8, at the end of a busy fiesta week, Santa Feans and visitors filled the Saint Francis Auditorium to honor the institutions' greatest patron.

As master of ceremonies, Hewett called forth a succession of speakers, each paying tribute to Springer and emphasizing a particular attribute of his life. Hewett commented on Springer's generosity, telling the audience, "If we could assemble here the young artists, musicians, and scientists who have been helped to their higher opportunities by him, the number would astonish you. And a greater effect than the material aid afforded was always the determined effort inspired by his faith. It called out the best in you." Hewett then told of the desire of Springer's friends to provide some permanent recognition of his contributions. With that he unveiled the bronze sculpture, which was accepted on behalf of the state by U.S. senator Andrieus A. Jones.[16]

Springer's role in the affair is best described by his son Wallace, who had driven to Santa Fe along with other friends from Colfax County. He described the occasion for his sister, Ada: "Pater was not present, and during a musical number he was sent for by a committee who escorted him in. All rose and broke forth in wildly enthusiastic applause. I was sitting at the opposite side of the room toward the front where my view was good, and I assure you it was one of the great moments of Pater's life, as well as being most impressive to me. As he walked in there was on his face an expression of modest satisfaction of one whose accomplishments have been recognized by friends and admirers. When the demonstration subsided, he spoke in a humorous vein, which was just the thing to relieve the tension. Everyone stood while he spoke. He wound up by dragging forth Scarpitta, the sculptor, for exhibition."[17]

Springer's remarks were characteristically modest and uncharacteristically brief, but he gave those in attendance to know that he had been deeply touched by the gesture. The recognition was well timed, as Springer's fragile health was soon to become a factor in his ability to carry on the active life to which he had been accustomed.

A Turning of Seasons

At the age of seventy-four, Frank Springer could look back on an era of extraordinary opportunity and mind-boggling change. Having come to New Mexico on the bone-jarring seat of a spring wagon, he had seen the territory remodeled to more nearly suit the values and sensibilities of ambitious newcomers like himself. He saw open prairie traversed for the first time by railroad grade, and he had taken part in the transformation that followed. He had traveled abroad on long ocean voyages to confer with men in Bonn and London and Amsterdam, and he had seen great distances shrink before the advance of aviation. He had lived boldly, had won a moderate fortune, and had given generously of his wealth, both to causes and to individuals. He had been impressed with the natural beauty of his adopted land and had spent joyful hours in contemplation of its rich history and prehistory. But by 1923 Springer could see the lengthening shadows and know that the evening of his days was at hand.

Springer entered the new year carrying on a pattern of activities that had been customary since his relocation to Washington. On January 8 he was in Raton for the annual meeting of the Maxwell Land Grant Company trustees. The meetings had become somewhat perfunctory with the diminution of operations; the company's few remaining business affairs were ably handled by Jan van Houten. This meeting involved little more than the obligatory election of officers and a review of the past year's receipts and disbursements. Springer was again elected president.

Springer made his usual summer sojourn, participating in a joint meeting of boards of the school and museum on July 30, 1923. After the meeting, he presided at a memorial service for Alice Fletcher, who had died in Washington on April 6.

Near the end of August, Springer suffered a heart attack. He was staying at the home of Carlos Vierra, in which he retained an interest. Springer was confined to bed during the first six weeks following the attack. It was an indication of his weakened condition that his son Edward was allowed to answer certain letters on his behalf. On October 7 Edward wrote to acknowledge a recent letter from Charles Lummis, who was unaware of Springer's infirmity. He told Lummis, "As you probably are not aware of his illness, I will describe it briefly. Six weeks ago Thursday he was taken violently ill with a heart attack, the result of an attack of indigestion, coupled with a too strenuous desire to walk a certain amount each day—and probably other causes. His recovery is now assured unless some unfortunate complication sets in, and we hope to take him from Santa Fe to California early in December or sooner if possible." Lummis responded immediately, expressing his concern and telling Edward, "Of all the men in the world, he is one of those I value most highly. And he is one of those who could least be spared in the world—they don't make his kind any more."[18]

By November 11 Springer was still too weak to leave his bed but was able to write a few lines of his own. He was unusually sparing in his use of words, but took the opportunity to acknowledge correspondence from Charles Schuchert at Yale, telling Schuchert, "Your letter of Oct. 18 found me here in bed with a severe heart attack, which pretty near took me across the Styx."[19] Springer did not go to California in December, and he was not in attendance when the Maxwell trustees met in February 1924.

The correspondence with Lummis became a comparison of maladies. While Springer endured with a dilated heart, Lummis underwent surgery to have a cataract removed. In January, Springer sent regards to Lummis, reporting on his inactivity over the past five months and conveying the dire nature of his illness. "I was very near passing over and had my transportation all arranged," he wrote, "but Charon, the old boatman, went back on me, and didn't come."[20] Significantly, Springer was convinced that his heart was mending.

May found Springer still in Santa Fe, carrying on a lively correspondence and taking up scientific work as his strength allowed. In September, a full year after the attack, Springer was able to preside over a joint meeting of the museum regents and the board of managers of the School of American Research. The boards engaged in a congenial bit of micromanagement, rearranging furniture in the Governors' Palace to suit the sensibilities of certain members. The minutes record that "[o]n motion of Twitchell and Walter, the bust of Lew Wallace presented by Henry Wallace to the School and Museum together with the Wallace arm chair and other memorabilia, were authorized to be removed from the Rito de los Frijoles room to the east end of the Palace to be placed on the spot pointed out by Mr. Frank Springer as the rostrum of the old Senate Chamber from which Governor Wallace addressed the Senate."[21]

Acknowledging the seriousness of his condition, Springer agreed to go east for examination and treatment. Upon reaching Philadelphia he entered Jefferson Hospital. On October 2, he could report that he was being sent home after a stay of ten days, telling Charles Schuchert. "'Home,' for me here is the residence of my daughter, whose husband is a very fine physician in other lines, who is to watch me for some weeks and note results. All of which leaves the question of return to my office somewhat uncertain."[22]

At Ada's home, Springer took up residence in a second-story bedroom, resumed his crinoid studies, and settled into the life of a "persistent invalid." Though in absentia, he continued in service to the Maxwell Land Grant Company, the Museum of New Mexico, and the School of American Research, executing his duties from afar and conferring with van Houten and Hewett in person and by letter. Josie stayed in Washington but traveled to Philadelphia from time to time to visit her husband and their daughter Ada. Their eldest daughter, Laura, also lived in the city with her husband, the attorney John J. K. Caskie.

The losses Springer could count at the end of 1924 included financial setbacks as well as physical ailments. The Eagle Nest project had failed to produce anticipated profits and the cattle market was in decline. Bank stock valued at twenty-eight thousand dollars in 1915 was rendered worthless when the San Miguel National Bank collapsed in 1924, along with other banks in Las Vegas and Santa Fe. Despite the reversal, Springer still had a substantial fortune, and he had no real financial worries as he settled into a quiet life in his daughter's home.

Charon, the "old boatman" to whom Springer had referred in remarks to Charles F. Lummis, was a figure from Greek mythology, a bad-tempered character who ferried the dead across the Styx, a river said to encircle the underworld of Hades. This was to become the persistent metaphor through which Springer spoke of his own mortality; each new health crisis became a tussle between Springer, the struggling mortal, and old Charon.

Following his relocation to Philadelphia, Springer spent another three years "grappling with the boatman," eluding Charon's hook and remaining among the living. His chief assets—his intellectual powers, his eternal optimism, and an indispensable sense of humor—never deserted him as he worked ahead in the upper room on Fifty-ninth Street, with occasional time out to do battle with his heart ailment. As a scientist and man of affairs, Springer had attained all that anyone might aspire to, but despite his infirmities and a permanent state of confinement, he remained a productive and useful citizen, as fully engaged with the world as his condition would allow.

CHAPTER 15

Across the Styx

Situated in his new surroundings—an upstairs bedroom in the comfortable suburban home of Warren and Ada Springer Davis—Springer made the best of a lifestyle that he would not have chosen, enduring without serious complaint the restrictions on his liberty and the occasional flare-ups of his chronic illness. He could hardly have had better care. Ada was a capable and attentive mistress of her home. A microbiologist with a master's degree from Columbia, she had assisted her father in his crinoid studies prior to her marriage and could appreciate the nature and value of his work. In Warren B. Davis, Springer had the daily care of an eminent physician who carried on a surgical practice and taught in the city's medical schools. Though not a cardiologist, he was well able to attend to the day-to-day needs of his ailing father-in-law.

While Springer became a permanent resident in his daughter's home, Josephine and Eva continued to live in Washington. Their travels now included regular visits to Philadelphia. The arrangement may have been based solely on Springer's need for medical attention, but one of the young grandsons, accustomed to being shushed and told, "We mustn't disturb Pater," assumed that his grandmother did not care to be with her husband on a regular basis.

Springer apparently never considered resigning his positions of responsibility in the Maxwell Land Grant Company, the Museum of New Mexico, and the School of American Research, but continued lending his good name, wise counsel, and influence, while others conducted meetings and carried out other routine functions. He was grateful for the fossil work that filled his days when his strength allowed and for borrowed time that enabled him to complete the research that he had blocked out as his contribution to science.

Frank Springer at the Davis Home, Overbrook, Pennsylvania, c. 1925. Courtesy of CS Cattle Company.

During the years of otherwise peaceful seclusion, Springer was forced to deal with a matter he had thought long settled. In March, 1925, U.S. senator William King of Utah wrote to inquire on behalf of the parents of Donald Beauregard, law clients of the senator's brother, Samuel King, concerning the disposition of Beauregard's works following his death in 1914. The paintings and drawings had been conveyed to Springer in consideration for the support given Beauregard for study in Europe. Springer had, in turn, given the works to the Museum of Fine Arts, with the stipulation that a selection be placed on exhibit in a room established in the young artist's memory. The family had seemed pleased with the arrangement, but Springer was now faced with the veiled suggestion that he had acted improperly, depriving the family of a valuable asset. The possibility of legal action was intimated.

Whenever Springer felt that his integrity was being questioned, he was apt to lead with his temper. His initial response to William King was courteous and informative, but upon learning that brother Samuel had been investigating the matter in Santa Fe, he followed up with a second letter, offering the senator a generous piece of his mind and challenging Samuel King to proceed with his lawsuit "and see where he will land."[1] Thoroughly alienated, Senator King sent a final missive expressing his resentment of the hostile sentiments and otherwise washing his hands of the matter.

Springer gathered up evidence of ownership, including a statement signed by Beauregard conveying the paintings to Springer. He sent the documents to Hewett in Santa Fe, along with a memorandum documenting the gift of Beauregard's works to the Museum of Fine Arts.[2] Having advised museum officials of the possibility of litigation, Springer revised his earlier estimate and concluded that a suit for recovery of the paintings was unlikely.

Springer's illness spared him from what likely would have been a greater involvement in a rift between Hewett and two longtime associates, Kenneth Chapman and Jesse Nusbaum. While serving as a functionary for Hewett in the Museum of New Mexico, Kenneth Chapman had also carved out an entrepreneurial stake in the cultural landscape of Santa Fe, leading a small group of associates in establishing the Pueblo Pottery Fund, later known as the Indian Arts Fund. The organization's purpose was to preserve fine examples of the native potter's art, while making traditional design motifs available to contemporary Pueblo potters. Nusbaum, meanwhile, had become superintendent at the new national park at Mesa Verde in Colorado.

Disenchanted with Hewett's management of the Museum of New Mexico, Chapman took advantage of Hewett's apparent disinterest in overtures from John D. Rockefeller concerning a possible grant to the museum.[3] Chapman and Nusbaum commenced their own cultivation of Rockefeller, proposing the creation of an autonomous Museum and Laboratory of Anthropology. Well knowing that Springer would be sensitive to any overt act of insurrection, Chapman approached the subject with care, first writing to apprise him of Rockefeller's interest in the Indian Arts Fund, then forwarding a copy of the already-submitted proposal. Predictably, Springer was cool to the idea, expressing dismay at the notion of dividing local support between the established institutions and a rival organization.[4]

Notwithstanding Springer's objection, the Museum and Laboratory of Anthropology was incorporated in August 1927, and promptly received a substantial gift from the Rockefeller Foundation. Nusbaum resigned the superintendency at Mesa Verde to become director of the new institution, with Chapman as his chief assistant. The Rockefeller gift included funds for a building to be designed by John Gaw Meem. The institution merged with the Museum of New Mexico in 1947, following Hewett's death and a decline in funds available for operating support.

His Life's Work

Springer's most pressing item of unfinished business was the completion of the monograph on the Silurian crinoids of North America. The work had been taking shape over a period of more than twenty years. Intrigued with an apparent relationship between Silurian species from the Swedish island of Gotland and similar fossils collected in Tennessee, Springer had begun a systematic investigation of crinoids of the Silurian period. He worked on the project as time allowed and had Kenneth Chapman produce drawings for eventual use as plates.[5]

Springer had only recently turned his attention to the project when he was stricken in August 1923. Realizing the precarious state of his health, he began to view the work as a race against time. In April 1924, while recovering at Vierra's home in Santa Fe, Springer told F. A. Bather of his good fortune in having his crinoid work at hand. Other projects remained to be finished, but Springer could now devote a major portion of his time to the new work, telling Bather: "I am now working with Chapman upon (what I hope will be) the last drawings for my work on the Silurian Crinoids, which I have had in hand for more than 15 years and contains much new material obtained by my collectors. I am now hoping to be able to finish this work and see it published, and it will be the last extensive work that I shall undertake."[6]

Springer found it prudent to arrange for a colleague to complete the work in the event of his death or incapacity. After careful consideration, he settled on Winifred Goldring, a promising young crinoid researcher, and so advised Ray Bassler at the National Museum. It did not prove necessary to involve Goldring, but she later attained distinction, serving as state paleontologist for New York and as president of the Paleontological Society.[7]

From his small world in the Davis home, Springer completed the writing, saw to details of publication, and oversaw the distribution of copies to his professional correspondents. By December 1926, copies of the published work were in hand. Springer wrote to Paul A. F. Walter at the Museum of New Mexico to advise him of the book's impending arrival, adding: "It will be my last work, and I am especially gratified to bring it to publication because it involves much material that depended wholly on my personal knowledge acquired during about 20 years of desultory preparation, as you will see from the introduction. Two years ago, during a particularly bad spell that I had, I made tentative arrangements to have the work undertaken by someone else with the aid of my scattered notes (scarcely intelligible to others) and using the drawings which were mostly completed. Persistent pegging away has avoided that unsatisfactory resort, and now that I have the work in print I feel doubly thankful for the allotment of *vis medicatrix* that has been vouchsafed to me."[8]

The publication of *American Silurian Crinoids* was a satisfying conclusion to half a century of scientific study and marked the completion of what Springer considered to be his life's work. In the introduction Springer told of difficulties encountered in working through a chronic and sometimes debilitating illness. "That I have been able to bring the work to a conclusion is a result that was not expected," he wrote, "and for which I am profoundly thankful."

American Silurian Crinoids was a serious scientific treatise, the culmination of many years' effort on Springer's part. For obvious reasons, however, the work drew numerous expressions of admiration for Springer and his broader contributions. In a review for *Science,* Charles Schuchert characterized the volume as a "beautiful and monumental work," and declared of its author, "Would that the human world had more men like Frank Springer!"[9]

Terminating the Maxwell Trust

One final task fell to Springer in his role as president of the trustees of the Maxwell Land Grant Company. By the end of 1926, the protracted process of paying off the company's prior lien bonds was drawing to a close. By provision of the deed of trust, the Dutch directors would, upon payment of the bonds, have the option of terminating the trust and assuming direct control of the company's assets. In anticipation of this time, Springer, Jan van Houten, and the Dutch principals began planning for the next phase of the company's operation.

Having considered the matter, Springer wrote to van Houten on December 4, 1926, suggesting two possible courses of action. In a letter of four pages, Springer laid out both options in a comprehensive and cogent manner. The letter demonstrated, among other things, that Springer, at the age of seventy-eight and having been away from the daily practice of law for some twenty years, remained in full possession of his legal and rhetorical skills. In closing, Springer wrote, "After you have received from our constituents in Amsterdam an intimation of the method they wish to adopt for the new organization, and the answers to the questions I have asked, I shall be glad to prepare for your consideration and theirs, forms of the various proceedings and documents which I think would be advisable in order to ensure a result that will be proof against hostile criticism."[10]

Responding on behalf of the Dutch company, H. A. Hartogh wrote from Amsterdam to acknowledge receipt of Springer's analysis, adding, "I'm glad to see that your health still allows you to give us your valuable advises, especially as the American laws are nightmares to us."[11] Hartogh clarified the position of the Dutch, expressing a preference for the first option proposed by Springer. Following this course of action, the directors would invoke article 10 of the trust deed, resulting in termination of the trust and reversion of all assets to the parent company. Hartogh also accepted Springer's offer of assistance in preparing the needed documents. A self-proclaimed invalid for more than two years, he turned to the task.

On January 7, 1927, the annual meeting of the Maxwell trustees was held in Raton. Springer was absent, but was again elected president. Through the spring Springer kept up a running correspondence with Hartogh and van Houten, and with the local attorney for the American trustees, Edwin C. Crampton of Raton. By the end of May, Springer was able to provide specific advice on procedures required to terminate the trust and place the remaining assets in the hands of the Dutch owners.

As promised, Springer drew up a three-part indenture to be signed by officials of the Dutch company, representatives of a majority of income bondholders, and the American trustees. Upon retirement of the last of the prior lien bonds, the indenture was to be filed in New Mexico and in the Netherlands. Springer wrote out the indenture in longhand. Forwarded to Raton, it was finalized by the company's New Mexico attorney, and a copy was sent to Springer for his signature in June 1927.

On August 8, however, van Houten wrote to advise of Crampton's preoccupation with other matters and a consequent delay in completing all the documents necessary to carry the plan into effect. "Apparently, with the delay we have had," he wrote, "it will be early spring before this transaction can be completed; and I hope that by that time I will be able to raise the funds necessary from some source."[12] Van Houten's final comment was an apparent reference to the sum of approximately fifty-five thousand dollars to be paid on the prior lien bonds. With remaining assets of approximately eight hundred thousand reported less than a year before, there should have been little doubt of the company's ability to raise the needed cash.[13]

The Boatman's Call

At this point, Springer's chronic illness intervened to prevent his seeing the trust termination through to its conclusion. Beginning in early June, his family had noticed a general decline in his condition.[14] At times he was able to function as before, but spells of weakness became longer and more frequent.

On August 3 Springer remained in good spirits, writing Edward a three-page letter, touching on a variety of matters including a lawsuit over water rights, recent rains at the family ranch, and the mischief of the range caterpillar. Springer found "quite exciting" the news of a recent visit by Vice Pres. Charles G. Dawes to Waite Phillips's ranch in Colfax County. Springer also mentioned a visit from his wife and daughter, writing that "Mater and Eva got away yesterday, as you are probably hearing direct from them—for Los Angeles via the Union Pacific." As for himself, he reported somewhat optimistically, "I am well again, and things moving here as usual."[15]

On August 21, however, Springer was unable to leave his bed, and his illness took on a more serious aspect. As Ada later recalled, "Various complications set in—kidney and stomach principally, which so weakened him generally that he was unable to gather enough strength to rally."[16] For some three weeks, the family held out hope that Springer would stage yet another of his remarkable recoveries, but all efforts at treatment failed. According to Ada's recollection:

> On the morning of September the thirteenth my father conversed with me at considerable length on subjects intricate and detailed in nature—He realized fully that his condition was grave. When he finished he said to me, "I am very tired and want to sleep." Within an hour he lapsed into a stupor from which he never moved except for a few brief seconds at a time.
>
> Unable to take nourishment, at the end not even water, he gradually wasted away. Strength failed him in every respect with the exception of his heart, which was strong and regular to the very last. One of the things hard to comprehend is this fact. For years his heart condition undermined his constitution, but at the last it was his strongest support.[17]

The boatman's final call came at two o'clock in the afternoon on Thursday, September 22, 1927. With his family gathered in the home where he had spent his last years, Frank Springer died quietly. At his request, his remains were cremated and the ashes were later taken to Las Vegas for burial. Private services were held in Philadelphia on Saturday, September 24.

Tributes

On September 23, nearly a full page of the *Santa Fe New Mexican* was devoted to an obituary written by Paul A. F. Walter. Said the *New Mexican* of Springer in an accompanying editorial, "Those who have known him entertain chiefly a lively recollection of a quiet, courteous, unselfish, kindly, considerate, modest gentleman. Mere casual introduction to him left this impression, strengthened steadily by successive contacts. Character is more immortal than museums or deserts which have been made to blossom, and we feel that the foregoing statement is as high [a] tribute as any man could covet."

Walter and others organized a memorial service in Santa Fe, and on the afternoon of Sunday, October 9, admirers filled the Saint Francis Auditorium. Present were persons connected with the school and museum, but also in attendance were attorneys and jurists, scientific colleagues, and people with whom Springer had been involved in business and civic matters. W. C. Reid, who had begun his career as a young attorney in Springer's office, characterized his former mentor as a solitary man. "He was not a man of the world, not a mixer," Reid said, "but those who knew him had nothing but the highest respect for him." The essence of Springer's career and civic involvement, said Reid, was that "[a]ll his work in life was basic and fundamental." This basic work included efforts to clear up the confusion of land titles in New Mexico; his commitment to improving education; his pursuit of science for its own sake; and his efforts to preserve the region's cultural wealth.[18]

Federal judge Orin L. Phillips elaborated on Springer's career and drew attention to one of the factors contributing to his success. "Mr. Springer was a man of almost inexhaustible energy and tremendous capacity for work," Phillips told the gathering. "Throughout his life he was a steady and indefatigable worker. Those who imagine that talent or genius can supply the place or achieve the results of labor, will find no encouragement in the life of Frank Springer."[19] Another service was held on October 12 in the Ilfeld Auditorium at Las Vegas, with friends again eulogizing Springer and emphasizing his devotion to the university.

Josephine and the Springer children received many expressions of sympathy, none more heartfelt than the message that came from Kenneth Chapman in Santa Fe. "It will take some time to fully realize that he has left us," Chapman wrote. "My work for him filled such a large part of my life. For anyone else, it would have been nothing more than a means of making a living, but his lovable personality made it a privilege, and I regretted more than I could possibly make him understand when it finally came to an end."[20]

Though others shared Chapman's sense of loss, there was also a shared if mostly unspoken understanding among Springer's family and friends that his had been a full and complete life, and that his last months and years had been, for them as for him, something of an unexpected coda. If Springer's spirit had forever gone, as he would have it, beyond the Stygian river, much remained to remind other mortals that he had passed this way.

CHAPTER 16

Reflections on a Many-Sided Life

On October 15, 1927, the Maxwell trustees met to consider the prospect of corporate life without the leader whose calm and steady judgment had long engendered confidence on the part of the trustees and their associates in the Netherlands. In truth, Jan van Houten had been carrying much of the load for some time. The process of terminating the trust was well along, and the Dutch would soon be in direct control of the company's affairs, but some functions remained to be fulfilled by the trustees.

Respecting the company's recent loss, van Houten told his colleagues, "The death of Mr. Springer is not only a great personal loss to all of us, but deprives the Board of his valuable counsel and advice. He was always cheerful and confident, and with his great mind and fine knowledge of law, was always able to analyze and decide the difficult questions of which so many presented themselves in the past years."[1]

In accordance with Springer's request, his son Wallace was elected to succeed him on the board. It must have come as a rude shock to the board and the family when, in March 1928, Wallace died in the crash of a light plane bound for Mexico City. Wallace had seemed a likely successor to his father as a business and civic leader in New Mexico, but this role now fell to Edward.

The termination of the Maxwell Trust proceeded much as anticipated, and by the end of 1928 the American board had been dissolved, leaving the Dutch firmly in control of the company. Jan van Houten continued in association with the Maxwell Land Grant Company until his death in 1949, persisting in mostly futile efforts to revive Baldy mining and overseeing the ongoing liquidation of the company's real property. The company survived to the end of the twentieth century as De Maxwell Petroleum Holding N.V., before a buyout in 2000 wrote an end to its long and eventful history.

Legacy

Frank Springer left an estate valued at more than half a million dollars. He had diversified his holdings, so that his assets consisted of common stocks, bonds, notes, real estate, life insurance, and cash. His largest investment, amounting to nearly half the estate, was in the Charles Springer Cattle Company. More than a quarter of the estate was in securities of the St. Louis, Rocky Mountain and Pacific Company.[2]

Following its formative years, the St. Louis, Rocky Mountain and Pacific Company assumed somewhat the character of a family business, with Charles Springer, Jan van Houten, and Edward T. Springer presiding over the company through more than forty years of its fifty-year history. The business prospered for several decades, but by 1950 the industry was in a slump and operations in Colfax County suffered accordingly. The St. Louis, Rocky Mountain and Pacific Company was sold to Kaiser Steel for $3.5 million in 1955. In relinquishing control of the company, Ed Springer recalled his father's conviction concerning the region's potential for coal production and expressed satisfaction that his vision had been realized.[3]

The Charles Springer Cattle Company, better known in recent years as the CS Cattle Company, was a family enterprise transcending the personal business interests of its namesake. Charles managed the ranch until 1908, when that responsibility was turned over to his nephew Hank Springer. Charles turned his attention to other interests, including the coal company, construction of the Eagle Nest Dam, and an increasing involvement in affairs of the New Mexico Republican Party. He died in 1932. After Hank's death, Ed Springer became general manager of the cattle company.

Charles Springer had no heirs, nor did any of Frank Springer's sons, but one of Ada's sons, James Leslie Davis, found the western country to his liking. Reared in Philadelphia, he entered Dartmouth College intending to become a physician. In 1941, however, he wrote to his Uncle Ed to propose that he come west and learn the cattle business. His education as a rancher was interrupted by the attack on Pearl Harbor and a call to military service, but he returned to the ranch a decorated veteran and assumed management of the CS Cattle Company in 1946. In 1953 he married Linda Mitchell,

James Leslie Davis, son of Ada Springer Davis, with his wife, Linda Mitchell Davis, at the CS Ranch, 1965. Courtesy of CS Cattle Company.

daughter of a leading New Mexico rancher, Albert K. Mitchell. When Les Davis died on May 12, 2001, he left a diversified ranching operation spread over more than 130,000 deeded acres and additional leased property. His six children had grown up to assume management of the cattle business, the lake and irrigation system, a growing business in game hunting, and a still-substantial horse breeding operation. In 2002 the CS disposed of a significant asset, selling the Eagle Nest Dam and Lake to the State of Mexico for $20 million.

Most of the cultural institutions established by Frank Springer and his contemporaries continue to thrive in the twenty-first century, if in somewhat altered form. New Mexico Highlands University is not the institution that Springer and Hewett envisioned—a southwestern variant of the elite universities with which they were familiar in other states. Reoriented to more nearly reflect the interests of the region's Hispanic population, the one-time normal university offers a diverse array of degrees and programs. The Ilfeld Auditorium, renovated in the mid-1990s, stands as the most visible vestige of the school's founding era.

It was perhaps inevitable that the interlocking relationship of the School of American Research and the Museum of New Mexico, as administered by Hewett, would not endure indefinitely. The role and influence of the school declined after Hewett's death in 1946, and a formal separation was concluded in 1959. Since that time the School of American Research has functioned as a private entity, eventually occupying an attractive campus in Santa Fe and sponsoring programs of advanced study, preservation, field research, publication, and public education. Meanwhile, the Museum of New Mexico has become more firmly established as the institutional base of a thriving tourist economy. A Museum of Indian Arts and Culture and a Museum of International Folk Art have been added, and facilities established by Springer and Hewett have undergone significant expansion.

Hewett remains an icon in the history of the School of American Research and the Museum of New Mexico. Springer's role in their development is less frequently mentioned, but his influence can hardly be overstated. Hewett's weaknesses were his patron's strengths. Springer forged useful alliances in the community, coaxed support from wary legislators, and more than once sustained Hewett when his survival as director seemed to hang in the balance. His material contributions were substantial and varied, and included commissioned mural paintings in the Palace and the Saint Francis Auditorium, the Finck linguistic library, works of art for the museum collection, subsidies for museum staff salaries, funds for construction of the Museum of Fine Arts, and a director's residence. Springer's financial, political, and moral support allowed the museum to grow rapidly and enabled Hewett to make a far better showing than would have been possible on the strength of government and AIA support alone.

Springer's legacy is also visible in the general acceptance of Pueblo-inspired architectural styles in Santa Fe and in the continued importance of archeological sites as regional assets. In the view of at least one observer, Springer was an influential proponent of the "western idea," publicly embracing a regional heritage based on the area's prehistoric cultures. His advocacy of these ideas helped counter the influence of Hewett's rivals, who were eager to haul artifacts back to eastern laboratories and museums, catalog them, and write papers about them, little caring about their value to contemporary residents of the region.[4]

Springer's crinoid collection and scientific library are intact, as he intended, in the National Museum of Natural History in Washington. Largely self-taught, Springer won the acclaim of accomplished scientists in the United States and Europe. While superseded to some extent, his work is still respected among contemporary researchers. As editor of volumes on the Crinoidea in the massive *Treatise on Invertebrate Paleontology,* Raymond C. Moore wrote, "It is reasonable to judge that no one in the world excels Frank Springer in adding to the knowledge of fossil crinoids." In the 1999 publication *Fossil Crinoids,* William Ausich described Springer as "the foremost North American crinoid worker" of his time. Of Springer's ultimate contribution, Ausich says, "He did much to comprehensively describe North American crinoids, and he provided much of the fundamental morphological, stratigraphic and systematic knowledge on which our current understanding of crinoids is based."[5]

Who Was Frank Springer?

Shortly after Springer's death, Ada Springer Davis was asked for her insight into her father's character. The question prompted thoughtful reflection.

> As I think back and recall some of the comments I have heard, my father was always somewhat of a puzzle to his friends, and not infrequently to the members of his own family. To many people he had—what we call out West—a poker face. By this I mean what thoughts or contemplated

> action might have been in his mind were difficult of interpretation. By nature he was very retiring—a man of few words—but his influence seemed always to be strongly felt. It was always difficult for me to reconcile his various interests and activities. His life on the frontier during the early seventies filled with thrilling adventure and romance—followed by a period of intense concentration on the practice of his profession, and lastly, the twenty-five years or more spent for the most part quietly on his scientific studies, were all so different.[6]

Frank Springer witnessed tremendous changes in a lifetime that began before the Civil War and saw the conquest and settlement of the West by traders, soldiers, farmers, ranchers, railroad builders, and capitalist merchants and developers. He took part in the contest for ownership of western lands and in the development of water, land, timber, and mineral resources. He experienced the rise of corporate America, endured a World War in which two sons fought, and lived to see his adopted territory become a state. He contributed materially to the development of the state's educational system and took a large hand in efforts to capitalize on its cultural resources.

Given the unusual range of Springer's activities and the routines that he adopted—leaving his family for weeks on end to conduct business or engage in solitary research; practicing law by day and studying extinct lifeforms by night; alternately entering into an active public life and retreating to a private world of scientific study—it is not surprising that friends and family members found him inscrutable. Even so, there are strong indications of the values and motives underlying his efforts.

The foundation of Springer's personal philosophy was his belief in science as the way of truth. One of the speakers who eulogized him in Santa Fe felt compelled to speculate that since his early days as an "infidel," Springer had adopted more conventional religious beliefs. The evidence that he offered was slight and unconvincing. In voluminous writings and public utterances, Springer occasionally referred to "Providence" or "the Powers of the Universe," but he never embraced a creed or affiliated with a religious body. At no time did Springer express hostility to conventional religious traditions, but his allegiance to science led to a rejection of doctrines that could not be substantiated by proof and precluded devotion to a belief system mediated by fellow mortals.

Springer's adult life coincided with two notable periods in the history of the United States, the Gilded Age of the late nineteenth century and the

Progressive era that was to follow. He rode the wave of Gilded Age fascination with industry and enterprise to professional and financial success, benefiting from a benevolent disposition on the part of Congress and the courts and adopting values that were prevalent among political and business leaders. At its extreme, the Gilded Age was a time of ostentation, greed, political and corporate corruption, and unrestrained pursuit of wealth. For men of more moderate appetites, for whom civic spirit and the rule of law imposed limits on personal ambition, the values of the age normalized and justified the notion of economic progress as the signal activity of civilized society and pointed to affluence as the measure of success.

The influence of the Gilded Age is evident in Springer's stewardship of the Maxwell Land Grant Company and its assets. While associated with the Maxwell Company, Springer was involved in numerous transactions that might in later years have been criticized on the basis of an apparent conflict of interest. While serving as legal counsel or president, he was involved in several ventures relying on major assets acquired from the company. These included not only the ranching enterprise in which he and Charles were partners, but the Springer Cement Company, the Maxwell Cattle Company, the St. Louis, Rocky Mountain and Pacific Company, the Continental Tie and Lumber Company, and the Cimarron Valley Land Company.

In another time, attentive shareholders might have asked how Springer could adequately safeguard the interests of the Maxwell Company while doing business with an entity in which he held a distinct and competing interest, but if the Dutch principals entertained such reservations they kept them to themselves. On the contrary, Springer was consistently praised by controlling members of the Dutch committee, whose confidence he had won through his success in the title suit and his competent administration of company affairs.

In view of the prevailing norms, and given the example of his father's contemporaries who eagerly occupied lands vacated by native tribes in Iowa, Springer could accept as normal the displacement of Indians, Hispanic farmers, and non-Hispanic squatters as an essential requisite to development of the Maxwell estate. His initial response to the Indian and Hispanic people he encountered in New Mexico was typical among nineteenth-century Anglo visitors and immigrants. He saw the Indians of Colfax County as a nuisance and an impediment to progress and fell in with other late arrivals in desiring their removal. Likewise, he was initially put off by the ubiquitous "mud houses," and by the rumored barbarism of the Penitentes.

Springer eventually found inspiration in native architecture and prehistoric cultures, but in 1917 he could not summon up much regret over the decline of the Anasazi, telling an audience at the dedication of the Museum of Fine Arts, "We cannot restore the original American to his lost dominion; nor from our view of human progress might it be desirable to do so."[7] He did support the appropriation of ancient cultures as objects of study and as the basis of a regional identity, and he embraced New Mexico's Hispanic heritage as a source of historical and romantic interest. He valued as colleagues men like Francisco Manzanares and J. Francisco Chávez, who had demonstrated success in business and politics.

By the turn of the century a strong Progressive movement was under way, giving impetus to political and social reforms. These reforms provided protections for workers, curbed excessive practices of corporations, and gave citizens greater access to democratic process through the extension of such procedures as the referendum, initiative, and recall. Springer's accomplishments in the territorial council of 1901, including substantial revisions to the public school laws and passage of an act to protect children from the evils of alcohol and tobacco, were consistent with the direction of Progressive reforms. When, in 1911, he unsuccessfully opposed the Blue Ballot amendment allowing citizens to more easily amend the new state constitution, Springer went against the tide of Progressive sentiment.

In view of his many interests and talents, Springer could fairly be characterized as a Renaissance man, but in disposition, he was a creature of the Gilded Age. Deliberate, controlled, and self-assured, he was generous with those who had worth in his scheme of values, and he sought to advance his interests and those of his clients within the parameters of the legal system. He was not visibly sympathetic toward the struggling men and women whose lives, in some instances, were made more difficult by his victories on behalf of the Maxwell Company and other clients. He accepted the notion of male dominance in business, politics, and family matters, yet late in life, he took pride in Eva's artistic career, supported Ada's pursuit of advanced studies, and exhibited high regard for female colleagues in science.

Frank Springer's example underscores the hazard inherent in summing up a life and then labeling a historic figure as a hero, a robber baron, a reformer, a revolutionary, or anything else. Springer was a man of varied interests and complex motives, and the impact of his life was equally diffuse. Said Shakespeare's Hamlet of his dead father, the king of Denmark, "He was a man, take him for all in all. I shall not look upon his like again."

As much may be said for Frank Springer. He was a force in New Mexico, in his profession, and in his scientific discipline, and none are apt to see his like again.

Despite shortcomings with respect to twenty-first-century standards of gender equity and social justice, Springer was wise in ways that enabled him to achieve extraordinary success. His view of the world took in eons of geologic history and encompassed ancient and extant civilizations on several continents. His intellect, energy, and work ethic propelled him to prominence in his professional life, and his financial success supported him in a variety of philanthropic, scientific, and cultural activities. A quiet but articulate man of action, he contributed substantially to the advancement of scientific knowledge and he left an indelible impression on the region that he claimed as his home, and in which for more than half a century he invested his vital energy.

Notes

Abbreviations

AA	Alexander Agassiz Papers
CFL	Charles Fletcher Lummis Papers
CSP	Charles Schuchert Papers
DB	Donald Beauregard Papers
ELH	Edgar Lee Hewett Papers
FS/FACHL	Frank Springer Papers, Fray Angélico Chávez History Library
FS/CSWR	Frank Springer Papers, Center for Southwest Research
FS/CS	Frank Springer Papers, CS Cattle Company
FS/NMNH	Frank Springer Papers, National Museum of Natural History
JLN	Jesse Logan Nusbaum Papers
KMC	Kenneth M. Chapman Papers
LW	Lew Wallace Collection
MAO	Miguel A. Otero Papers
MLG	Maxwell Land Grant Collection
NARA	National Archives and Records Administration
NHM	Natural History Museum, London
NMHU	New Mexico Highlands University
PAFW	Paul A. F. Walter Papers
RG	Record Group
RSB	Ray S. Bassler Papers

Chapter 1

1. William Salter, *Iowa: The First Free State in the Louisiana Purchase,* 271.

2. Francis Springer, "Recollections of Judge Francis Springer," *Annals of Iowa* 2 (Jan., 1897): 572. Much of the information in this section is owed to this source.

3. D. N. Sprague, "Pioneer Statesmen: A Gentleman of the Old School," *Columbus (Iowa) Safeguard,* Feb. 7, 1896.

4. *Santa Fe New Mexican,* Feb. 9, 1901.

5. Margaret N. Keyes, *Old Capitol: Portrait of an Iowa Landmark,* 23–26.

6. Frank Springer to Nancy R. Springer, March 11, 1864, FS/CS.

7. Pocket diary of Frank Springer, 1865, Apr. 15, 1865, FS/CS.

8. Frank Springer to Forest C. Ensign, Jan. 14, 1913, FS/NMNH.

9. Frank Springer to Nancy R. Springer, Nov. 21, 1863, FS/CS.

10. Pocket diary of Frank Springer, 1867, Feb. 9, 1867, FS/CS.

11. Frank Springer to Nancy R. Springer, Feb. 24, 1867, FS/CS.

12. Frank Springer, "Science as a Civilizer," June, 1867, typescript, FS/FACHL.

13. Frank Springer to Nancy R. Springer, March 13, 1870, FS/CS.

14. See "The Law Department of Iowa State University," *University (of Iowa) Reporter,* July, 1870.

15. Frank Springer, "Memorandum Furnished at Request of R. E. Twitchell," manuscript, n.d., FS/CS.

16. Pocket diary of Frank Springer, 1869, Aug. 5–9, 1869.

17. Springer, "Memorandum Furnished at Request of R. E. Twitchell," n.d., FS/CS.

18. Ibid.

19. *Burlington (Iowa) Daily Gazette,* Apr. 22, 1871.

20. *The State v. Lawrence,* 38 *Iowa Reports* 51 (1873).

21. R. E. Twitchell, *Notes and Memoranda Concerning the Legal, Literary, and Scientific Activities of Mr. Frank Springer as a Citizen of New Mexico,* 13.

Chapter 2

1. See Norman Cleaveland, with George Fitzpatrick, *The Morleys: Young Upstarts on the Southwest Frontier,* for the most complete account of Morley's life and career.

2. For this account of the early history of the Maxwell Land Grant, the author is indebted to Jim Berry Pearson, *The Maxwell Land Grant.*

3. This is according to Porter's recollection in Henry M. Porter, *Pencilings of an Early Western Pioneer,* and Pearson's account in *The Maxwell Land Grant.*

4. Agreement between the Maxwell Land Grant and Railway Company and William R. Morley, with reference to engagement as Executive Officer of the Company, Nov. 11, 1872, FS/CS.

5. Pearson, *The Maxwell Land Grant,* 59–61.

6. N. Cleaveland, with Fitzpatrick, *The Morleys,* 72–73.

7. Morley to Stephen B. Elkins, Jan. 27, 1873, MLG.

8. Frank Springer to Nancy R. Springer, Jan. 29, 1873, FS/CS.

9. Frank Springer to Francis Springer, Jan. 31, 1873, FS/CS.

10. Frank Springer to Nancy R. Springer, Feb. 7, 1873, FS/CS.

11. Ibid.

12. Morley to Thomas B. Catron, Feb. 7, 1873, FS/CS.

13. Morley to Catron, Feb. 11, 1873, FS/CS.

14. Frank Springer to Francis Springer, March 4, 1873, FS/CS.

15. Frank Springer to Charles Wachsmuth, Apr. 23, 1873, FS/NMNH.

16. Springer to Wachsmuth, July 15, 1873, FS/NMNH.

17. Frank Springer to Francis Springer, July 27, 1873, FS/CS.

18. Frank Springer to Francis Springer, Nov. 17, 1873, FS/CS.

19. Stephen B. Elkins to W. R. Morley, March 17, 1874, FS/CS.

20. Ibid. Along with Catron, Henry L. Waldo of Santa Fe handled some legal matters for the Maxwell Land Grant and Railway Company at Elkins's direction.

21. *Cimarron (N.Mex.) News,* Jan. 17, 1874.

22. Veronica E. Velarde Tiller, *The Jicarilla Apache Tribe,* 77–98.

23. *Cimarron (N.Mex.) News,* June 13, 1874.

24. Springer to Wachsmuth, Jan. 8, 1875, FS/NMNH.

25. Ibid.

26. *Santa Fe Weekly New Mexican,* Nov. 24, 1874.

27. Agreement concluded by letter from W. R. Morley to Will D. Dawson, Dec. 3, 1874, accepted by Dawson on Dec. 7, 1874, FS/CS.

28. *Elizabeth City Railway, Press, and Telegraph,* Apr. 18, 1874.

29. *Cimarron (N.Mex.) News,* June 20, 1874.

30. Morris F. Taylor, *O. P. McMains and the Maxwell Land Grant Conflict,* 36.

31. *Cimarron (N.Mex.) News and Press,* May 29, 1875.

32. On the occasion of the graduation of the 1869 class in law at the State University of Iowa, witnessed by Springer, the same "L.E.X." contributed an account of the proceedings to Springer's hometown paper, the *Burlington (Iowa) Hawk-Eye.* The command of legal knowledge and rhetoric in the *Cimarron (N.Mex.) News and Press* article also points to Springer.

33. *Cimarron (N.Mex.) News and Press,* Aug. 28, 1875.

34. Victor Westphall, *Thomas Benton Catron and His Era,* 115.

35. *Santa Fe Daily New Mexican,* Sept. 13, 1875.

36. *Santa Fe Daily New Mexican,* July 30, 1875.

37. *New York Sun,* July 5, 1875.

38. Deposition of Frank Springer, of Cimarron, N.Mex., given at the request of Frank Warner Angel, Esq., upon interrogations propounded by him, Aug. 9, 1878, *Report and Testimony in the Matter of Charges against Samuel B. Axtell, Governor of New Mexico,* Oct. 3, 1878, NARA, RG 48.

39. N. Cleaveland, with Fitzpatrick, *The Morleys,* 90.

40. Melvin Mills to Robert Longwill, July 18, 1875. A copy of the letter is included in Frank Springer's deposition of Aug. 9, 1878, NARA, RG 48.

Chapter 3

1. Accounts of events known generally as the "Colfax County War" may be found in N. Cleaveland, with Fitzpatrick, *The Morleys;* Lawrence R. Murphy, *Philmont: A History of New Mexico's Cimarron Country;* and Taylor, *O. P. McMains.* María Montoya provides a somewhat different perspective in *Translating Property: The Maxwell Land Grant and the Conflict over Land in the American West, 1840–1900.*

2. See *Las Vegas (N.Mex.) Daily Optic,* Sept. 2, 1884.

3. Westphall, *Thomas Benton Catron,* 103–108.

4. Deposition of Frank Springer, Aug. 9, 1878, *Report and Testimony in the Matter of Charges against Samuel B. Axtell,* NARA, RG 48.

5. Thomas Harwood, *History of New Mexico Spanish and English Missions of the Methodist Episcopal Church from 1850 to 1910,* 266.

6. This version of events, found in Taylor, *O. P. McMains,* 39–41, is based largely on a report published in the *Santa Fe Weekly New Mexican,* Sept. 4, 1877.

7. Testimony of Isaiah Rinehart at the trial of O. P. McMains for the murder of Cruz Vega, reported in the *Cimarron (N.Mex.) News and Press,* Aug. 30, 1877.

8. Deposition of Frank Springer, Aug. 9, 1878, NARA, RG 48.

9. Ibid.

10. Ibid.

11. *Santa Fe Weekly New Mexican,* Nov. 16, 1875.

12. Deposition of Frank Springer, Aug. 9, 1878, NARA, RG 48.

13. "Proceedings of the Citizens' Meetings," Nov. 10, 1875, NARA, RG 48.

14. Harwood, *History of New Mexico Spanish and English Missions,* 267.

15. *Cimarron (N.Mex.) News and Press,* Dec. 31, 1875.

16. *Cimarron (N.Mex.) News and Press,* Jan. 7, 1876.

17. *Cimarron (N.Mex.) News and Press,* Jan. 28, 1876.

18. Deposition of Frank Springer, Aug. 9, 1878, NARA, RG 48.

19. N. Cleaveland, with Fitzpatrick, *The Morleys,* 113.

20. S. B. Axtell to W. R. Morley, Jan. 20, 1876, FS/CS. The full exchange of telegrams between the Cimarron men and the governor is reported in N. Cleaveland, with Fitzpatrick, *The Morleys,* 113–14.

21. Deposition of Frank Springer, Aug. 9, 1878, NARA, RG 48.

22. This account of events is as related in Frank Springer's deposition of Aug. 9, 1878, as well as his affidavit of June 10, 1878, accompanying a letter to Carl Schurz, U.S. Secretary of the Interior, in which Springer brings charges against Gov. S. B. Axtell, NARA, RG 48.

23. Frank Springer, "Who's a Liar?: Mr. Springer's Respects to Governor Axtell, a Post Mortem on El Obispo," *Cimarron (N.Mex.) News and Press,* Oct. 3, 1878.

24. S. B. Axtell to O. K. Chittenden, Feb. 21, 1876, holographic copy, FS/CS.

25. O. K. Chittenden to Frank Springer, Sept. 26, 1878, FS/CS. "Judge Waldo" was Henry Waldo, then chief justice of the territory of New Mexico.

26. Brigham Young to Carl Schurz, July 13, 1877, NARA, RG 48.

27. The most detailed account of the prosecution of O. P. McMains is found in Taylor, *O. P. McMains,* 48–56. The present work leans heavily on Taylor's account.

28. *Cimarron (N.Mex.) News and Press,* Aug. 30, 1877.

29. *Colorado Chieftain,* Sept. 6, 1877.

30. *Cimarron (N.Mex.) News and Press,* Aug. 30, 1877.

31. Thomas Harwood, letter to the editor, *Las Vegas (N.Mex.) Daily Optic,* Feb. 10, 1890, reproduced in Harwood, *History of New Mexico Spanish and English Missions,* 272–75.

32. Affidavit of Frank Springer, Aug. 30, 1877, NARA, RG 48.

33. Deposition of Frank Springer, Aug. 9, 1878, NARA, RG 48.

34. *Cimarron (N.Mex.) News and Press,* Jan. 17, 1878.

35. As published in the *Cimarron (N.Mex.) News and Press,* Apr. 18, 1878.

36. The full text of letters exchanged between Springer and Axtell between April 4 and May 13, 1878, was published in the *Cimarron (N.Mex.) News and Press,* Oct. 3, 1878.

37. Frank Springer to Carl Schurz, June 10, 1878; affidavit of Frank Springer, June 10, 1878, NARA, RG 48.

38. N. Cleaveland, with Fitzpatrick, *The Morleys,* 152.

39. Westphall, *Thomas Benton Catron,* 121, 127–28.

40. George W. McCrary to Edward Hatch, Sept. 18, 1878, LW.

41. *Cimarron (N.Mex.) News and Press,* Sept. 6, 1878.

42. *Santa Fe Weekly New Mexican,* Sept. 21, 1878; Oct. 19, 1878.

43. *Cimarron (N.Mex.) News and Press,* Oct. 3, 1878.

44. N. Cleaveland, with Fitzpatrick, *The Morleys,* 157–58.

45. Wallace's arrival in Santa Fe is described in full in a letter to his wife, Susan, Oct. 8, 1878, LW.

46. *Cimarron (N.Mex.) News and Press,* Oct. 3, 1878.

47. Lew Wallace to Carl Schurz, Oct. 1, 1878, LW.

48. Frank Warner Angel, *Report and Testimony in the Matter of Charges against Samuel B. Axtell, Governor of New Mexico,* Oct. 3, 1878, NARA, RG 48.

49. See Pearson, *The Maxwell Land Grant,* 73–78. This brief account of the condition of the Maxwell Land Grant Company, 1875–78, is chiefly owed to Pearson.

50. Charles E. Gast to Committee of Share and Bondholders of the Maxwell Land Grant and Railway Company, Sept. 1, 1877, MLG.

Chapter 4

1. *Santa Fe Weekly New Mexican,* Oct. 19 and Nov. 2, 1878.

2. F. Stanley, *Clay Allison,* 107; N. Cleaveland, with Fitzpatrick, *The Morleys,* 83.

3. Dates and places of birth, dates of some deaths and marriages, and the date of the family's removal from Cimarron to Las Vegas, N.Mex., are found in the "Family Record," entered in Frank Springer's hand, FS/CS.

4. Frank Springer to Josephine Springer, Feb. 5, 1878, FS/CS.

5. Accounts of the AT&SF-D&RG rivalry, and of Morley's involvement, may be found in several histories of railroad construction in the West. Among these are Keith Bryant, *History of the Atchison, Topeka and Santa Fe Railway;* L. L. Waters, *Steel Trails to Santa Fe;* and for a somewhat more popular account, James Marshall, *Santa Fe: The Railroad That Built an Empire.*

6. *Trinidad (Colo.) Daily News,* Dec. 10, 1878.

7. *Cimarron (N.Mex.) News and Press,* Dec. 12, 1878.

8. Frank Springer to Josephine Springer, May 11, 1879, FS/CS.

9. Frank Springer to A. A. Robinson, Dec. 29, 1906, copy in files of Marc Simmons, private collection.

10. Ibid.

11. Frank Springer to Josephine Springer, June 4, 1879, FS/CS.

12. Frank Springer to Josephine Springer, July 23, 1879, FS/CS.

13. The Santa Fe was officially styled the Atchison, Topeka and Santa Fe Railroad from 1863 to 1895, when the name was changed to Atchison, Topeka and Santa Fe Railway.

14. Frank Springer to Josephine Springer, June 4, 1879, FS/CS.

15. Howard Roberts Lamar, *The Far Southwest, 1846–1912: A Territorial History,* 174.

16. *Santa Fe Sentinel,* July 24, 1879; Frank Springer to Josephine Springer, July 23, 1879, FS/CS.

17. Frank Springer to Josephine Springer, July 30, 1879, FS/CS.

18. Frank Springer to Josephine Springer, Jan. 11, 1880, FS/CS. "Young Romero" may have been Rafael Romero, who served as Speaker of the House of Representatives and represented Mora County, while Springer represented Colfax and Mora in the Council.

19. Frank Springer to Josephine Springer, Aug. 14, 1879, FS/CS.

20. Frank Springer to Josephine Springer, Sept. 15, 1879, FS/CS.

21. Frank Springer to Josephine Springer, Feb. 5, 1882, FS/CS.

22. Frank Springer to Josephine Springer, Aug. 20, 1882, FS/CS.

23. Frank Springer to Josephine Springer, Apr. 28, 1883, FS/CS.

24. Frank Springer to Josephine Springer, Sept. 4, 1883, FS/CS.

25. Frank Springer to Francis Springer, June 18, 1882, FS/CS.

26. Francis Springer to David Davis, June 24, 1882, copy, FS/CS.

27. Frank Springer to Josephine Springer, June 30, 1882, FS/CS.

28. Frank Springer to the Honorable Judiciary Committee of the Senate, July 3, 1882, copy, FS/CS.

29. W. D. Lee to Frank Springer, June 28, 1882, FS/CS.

30. *Santa Fe Daily New Mexican,* July 6, 1882.

31. *Santa Fe Daily New Mexican,* July 11, 1882.

32. Frank Springer to Josephine Springer, July 4, 1882, FS/CS.

33. Frank Springer to Josephine Springer, July 9, 1882, FS/CS.

34. Frank Springer to Josephine Springer, July 15, 1882, FS/CS.

35. *Santa Fe Daily New Mexican,* July 18, 1882.

36. *Santa Fe Daily New Mexican,* July 27, 1882.

37. *Santa Fe Daily New Mexican,* Aug. 1, 1882.

38. For a brief sketch of Axtell's life and career, see William A. Keleher, *Violence in Lincoln County, 1869–1881: A New Mexico Item,* 28–30.

39. Principals in the Maxwell Cattle Company were Sherwin, president; Frank Springer, vice president; Harry Whigham, secretary; H. M. Porter, treasurer; and Manly M. Chase, general manager. For an authoritative account of affairs of the enterprise, see Morris F. Taylor, "The Maxwell Cattle Company, 1881–1888."

40. See J. T. Nieuwenhuisen, "To the Readers of the *News and Press,*" July 20, 1881.

41. *Las Vegas (N.Mex.) Gazette,* Jan. 10, 1883.

Chapter 5

1. *Santa Fe Daily New Mexican,* Apr. 19, 1887. Quotes in this and the following paragraph are from this edition of the newspaper.

2. *Raton (N.Mex.) Guard,* Apr. 14, 1882.

3. Springer, "Land Titles in New Mexico," address before the New Mexico Bar Association. Also carried in the *Santa Fe Daily New Mexican,* Jan. 7, 1890.

4. For an extensive treatment of this idea, see María E. Montoya, *Translating Property.*

5. One of the best histories of Spanish and Mexican land grants in the Southwest may be found in Malcolm Ebright, *Land Grants and Lawsuits in Northern New Mexico,* 11–54. A good summary of the early legal history of the Maxwell Grant may be found in Taylor, *O. P. McMains,* 57–73. The summary provided in this chapter is largely based on these two accounts.

6. See Lawrence R. Murphy, *Lucien Bonaparte Maxwell: Napoleon of the Southwest.*

7. Willis Drummond to James K. Proudfit, Jan. 28, 1874, Records of the Surveyor General of the Territory of New Mexico.

8. Taylor, *O. P. McMains,* 63.

9. S. F. Phillips to Frank Springer, Nov. 19, 1881. Reproduced in full in "The Maxwell Land Grant: Opinion of the Attorney General of the United States, Declining to Bring Suit to Vacate the Patent, November 19, 1881," privately printed for Frank Springer, n.d., FS/CS.

10. S. J. Kirkwood to Frank Springer, Dec. 30, 1881, "The Maxwell Land Grant."

11. *Raton (N.Mex.) Guard,* March 24, 1882.

12. *Raton (N.Mex.) Comet,* July 14, 1882.

13. *Raton (N.Mex.) Comet,* July 28, 1882.

14. *United States v. Maxwell Land-Grant Co. et al.* 121 U.S. 325 (1887), transcript of record, Case File 12100, NARA, RG 267.

15. *Raton (N.Mex.) Comet,* Feb. 15, 1884.

16. *Raton (N.Mex.) Comet,* Feb. 22, 1884.

17. *Raton (N.Mex.) Comet,* March 7, 1884.

18. *Raton (N.Mex.) Comet,* March 14, 1884.

19. *United States v. Maxwell Land-Grant Co. and Others,* 21 Federal Reporter 19 (1884).

20. *Raton (N.Mex.) Comet,* Sept. 19, 1884.

21. *Raton (N.Mex.) Comet,* Oct. 10, 1884.

22. See Ebright, *Land Grants and Lawsuits,* 43–45, 110–11.

23. *United States v. Maxwell Land-Grant Company et al.,* 121 U.S. 325 (1887), transcript of record.

24. *Denver Tribune,* Nov. 6, 1885.

25. *United States v. Maxwell Land Grant Co. and others,* 26 Federal Reporter 118, (1886).

26. Stephen B. Elkins to Frank Springer, Feb. 9, 1886, FS/CS.

27. *Raton (N.Mex.) Weekly Independent,* March 20, 1886.

28. *Raton (N.Mex.) Comet,* March 26, 1886.

29. *Raton (N.Mex.) Weekly Independent,* Apr. 3, 1886.

30. *Raton (N.Mex.) Comet,* Apr. 9, 1886.

31. O. P. McMains, "Conspiracy to Defraud," broadside dated Apr. 24, 1886, Raton, N.Mex., FS/CS.

32. *Raton (N.Mex.) Comet,* May 14, 1886.

33. Twitchell, *Notes and Memoranda,* 19.

34. Ibid., 24.

35. Ibid., 24–25.

36. Henry Strong to Francis Springer, March 11, 1887, FS/CSWR.

37. Ibid.

38. *Santa Fe Daily New Mexican,* Apr. 21, 1887.

39. Samuel F. Miller to Francis Springer, Apr. 24, 1887, FS/CSWR.

40. *Raton (N.Mex.) Daily Range,* May 20, 1887.

41. *United States v. Maxwell Land-Grant Company et al.,* 122 U.S. 365 (1887).

42. Frank Springer to Charles Wachsmuth, May 22, 1887, FS/NMNH.

43. *United States v. Maxwell Land Grant Co.,* 5 N.Mex. 297 (1889).

44. *Raton (N.Mex.) Daily Range,* June 10, 1887.

45. *New York Times,* Oct. 12, 1887.

46. A good account of the "Stonewall War" may be found in Taylor, *O. P. McMains,* 211–25.

47. McMains, *Defiance,* tract, privately printed c. Oct., 1888.

48. Deed of Trust, Maxwell Land Grant Company, Jan. 19, 1888, MLG.

Chapter 6

1. Brief notes on the family's travels in California are contained in Frank Springer's pocket diaries for 1892 and 1894, FS/CS.

2. See Morris F. Taylor, "The Maxwell Cattle Company, 1881–1888," *New Mexico Historical Review* 49:4 (1974): 289–324.

3. Ibid., 293–95.

4. Ibid., 295. Taylor cites minutes of the Maxwell Land Grant Company board, Feb. 20, 1883, and a letter from Harry Whigham to Martinus Petrus Pels, dated Apr. 11, 1885, contained in letter book 3 of the Maxwell Land Grant Company records, MLG.

5. See Hans Mattson, *Reminiscences: The Story of an Immigrant,* 276–89.

6. In an 1884 recapitulation of the Maxwell Land Grant Company's debts, Mattson reckoned the obligation to Sherwin at $675,000. Victor van Lint, "The Maxwell Land Grant," typescript, n.d., MLG.

7. Frank Holmes, "The Horses of the CS Ranch," pt. 1, 161–62.

8. Porter, *Pencilings of an Early Western Pioneer,* 21–22.

9. Springer-Wachsmuth correspondence, FS/NMNH. Several letters dated in Fall, 1887, describe Springer's activities in Europe en route to and from Amsterdam.

10. Twitchell, *Notes and Memoranda,* 5–6.

11. Pearson, *The Maxwell Land Grant,* 111.

12. Ibid., 144–77.

13. For an explanation of the Beales claim, see Victor Westphall, *Mercedes Reales: Hispanic Land Grants of the Upper Rio Grande Region,* 37–40.

14. *Interstate Land Company v. Maxwell Land Grant Company et al.* 139 U.S. 569 (1891).

15. Minutes of the Executive Committee of the Board of Trustees, Maxwell Land Grant Company, Aug. 6, 1891, minute book of Jan. 1888 to Jan. 1893, MLG.

16. Springer to Charles Wachsmuth, Apr. 5, 1881, FS/NMNH.

17. This is Martinsen's account, contained in his letter to the American trustees, Oct. 2, 1891, MLG.

18. Ibid.

19. Frank Springer to S. van Houten, Sep. 10, 1891, MLG.

20. Frank Springer to M. P. Pels, Dec. 14, 1891, MLG.

21. Minutes of the Annual Meeting of the Board of Trustees, Maxwell Land Grant Company, Feb. 1, 1892, MLG.

22. *Memorandum.* Notes in Springer's hand indicate that the memorandum was distributed to Maxwell investors in Jan. and Feb., 1893, FS/CS.

23. Frank Springer to Charles Wachsmuth, March 13, 1895, FS/NMNH.

Chapter 7

1. Twitchell, *Notes and Memoranda,* 12.

2. Charles Rollin Keyes, "An Epoch in the History of American Science," 348.

3. Aldo Leopold, *A Sand County Almanac, with Other Essays on Conservation from Round River,* 170–71.

4. Though unnamed in the text, Wachsmuth is identified as the subject of Leopold's comments by Wayne I. Anderson in *Iowa's Geologic Past,* 218–19.

5. Much of the biographical information concerning Wachsmuth is owed to C. Keyes, "An Epoch in the History of American Science."

6. "C. Wachsmuth Dies," *Burlington (Iowa) Daily Gazette,* Feb. 11, 1896.

7. C. Keyes, "An Epoch in the History of American Science," 350.

8. Pocket diary of Frank Springer, 1869, March 28, 1869, FS/CS.

9. Field diary of Frank Springer, 1867–99, May 8, 1871, FS/NMNH.

10. C. Keyes, "An Epoch in the History of American Science," 351.

11. Springer to Wachsmuth, March 21, 1877, FS/NMNH.

12. Springer to Wachsmuth, Feb. 23, 1879, FS/NMNH.

13. Springer to Wachsmuth, May 30, 1879, FS/NMNH.

14. Springer to Wachsmuth, Nov. 30, 1880, FS/NMNH.

15. Springer to Wachsmuth, Apr. 5, 1881; Nov. 21, 1882, both in FS/NMNH.

16. Jan F. Van Sant and N. Gary Lane, *Echinodermata,* 20.

17. Springer to Wachsmuth, Jan. 9, 1887, FS/NMNH.

18. Springer to Agassiz, Feb. 15, 1888, AA.

19. Springer to Wachsmuth, May 29, 1893, FS/NMNH.

20. Springer to Agassiz, Dec. 21, 1893, AA.

21. Wachsmuth to Agassiz, Sept. 15, 1894, AA.

22. Springer to Wachsmuth, Oct. 31, 1894, FS/NMNH.

23. Springer to Agassiz, Oct. 31, 1894, AA.

24. Springer to Wachsmuth, Nov. 25, 1894, FS/NMNH.

25. Springer to Wachsmuth, Jan. 12, 1896, FS/NMNH.

26. "C. Wachsmuth Dead," *Burlington (Iowa) Daily Gazette,* Feb. 11, 1896.

27. C. Keyes, "An Epoch in the History of American Science."

28. "Note" in Wachsmuth and Springer, *Monograph of the North American Crinoidea Camerata,* 1897.

Chapter 8

1. Springer, "Memorandum Furnished at Request of R. E. Twitchell," n.d., FS/CS.

2. Typewritten note on stationery of the CS Cattle Company, FS/CS. These principles are also quoted in Twitchell, *Notes and Memoranda.*

3. Twitchell, *Notes and Memoranda,* 7.

4. Frank Springer, "The Bar and the People," speech to the New Mexico Bar Association, Jan. 4, 1900, published in *Santa Fe New Mexican,* Jan. 5, 1900; also in Twitchell, *Notes and Memoranda,* 58–60.

5. Springer, "Land Titles in New Mexico," 1890.

6. See Delphine Dawson Wilson, *John Barkley Dawson: Pioneer, Cattleman, Rancher.*

7. *Maxwell Land Grant Company v. John B. Dawson,* 7 N.Mex. 133 (1893). The blank line is in the original deed; no number is filled in indicating the number of acres.

8. D. Wilson, *John Barkley Dawson,* 83.

9. William A. Keleher, *Maxwell Land Grant: A New Mexico Item,* 143. Keleher refers to Springer as "Frank W. Springer." Others have also tried to give Springer a middle initial—E and L among them—but he had no middle name or initial.

10. Lansing B. Bloom, "Charles Springer," *Dictionary of American Biography,* 11:671.

11. *Cunningham v. Springer,* 13 N.Mex. 259 (1905).

12. The date of the conversation and agreement is according to Jones's representations, *Cunningham v. Springer,* 13 N.Mex. 259 (1905).

13. *Maxwell Land Grant Company v. John B. Dawson,* 7 N.Mex. 133 (1893).

14. *Springer Land Association v. Ford,* 168 U.S. 513 (1897).

15. *Thompson v. Maxwell Land-Grant and Railway Company,* 168 U.S. 451 (1897).

16. *Las Vegas (N.Mex.) Daily Optic,* Nov. 9, 1897.

17. Twitchell, *Notes and Memoranda,* 8.

18. Ibid., 10.

19. See Westphall, *Thomas Benton Catron,* 230–68.

20. W. C. Reid, "The Springer Memorial Services: Address by Captain W. C. Reid," *El Palacio* 23: 14–16, Oct. 22, 1927, 371.

Chapter 9

1. The most authoritative source on the early history of the university may be Hugh W. Price, "A History of the New Mexico Normal University, 1893–1931." The author is indebted to this source for much general information in this chapter.

2. *Record of Proceedings, Las Vegas Normal School,* March 1893 through July 15, 1915, NMHU.

3. Ibid., regents minutes of Oct. 4, 1897, NMHU.

4. Edgar L. Hewett, *Two Score Years,* 107.

5. Frank Springer, remarks at a dinner in honor of Charles Ilfeld, Apr. 14, 1919, in Price, "A History of the New Mexico Normal University," 84.

6. Price, "A History of the New Mexico Normal University," 15–16.

7. Frank Springer, "On the Dedication of the New Mexico Normal University, Las Vegas; March 4, 1899," in *Educational Addresses of Dr. Frank Springer,* 1927.

8. "Statistics 1898–1935," Ledger Book, NMHU.

9. Bulletin of the New Mexico Normal University, 1901, 1902, NMHU.

10. James Graham McNary, *This Is My Life,* 21–23.

11. Bulletin of the New Mexico Normal University, 1901, NMHU.

12. *Las Vegas (N.Mex.) Daily Optic,* Oct. 23, 1900.

13. *Las Vegas (N.Mex.) Daily Optic,* Nov. 5, 1900.

14. *Las Vegas (N.Mex.) Daily Optic,* Nov. 14, 1900.

15. *Santa Fe New Mexican,* Feb. 9, 1901.

16. *Las Vegas (N.Mex.) Daily Optic,* Feb. 8, 1901.

17. *Las Vegas (N.Mex.) Daily Optic,* March 16, 1901.

18. A. E. Winship, "New Mexico Annual Meeting," *New England Journal of Education,* Jan. 22, 1903.

19. *Record of Proceedings, Las Vegas Normal School,* regents minutes of Oct. 13, 1902, NMHU.

20. Ibid., regents minutes of Nov. 13, 1902.

21. Ibid., regents minutes of Jan. 15, 1903.

22. Ibid., regents minutes of Jan. 19, 1903.

23. Charles Ilfeld to Otero, Jan. 20, 1903, MAO.

24. Otero to Springer, Jan. 23, 1903, MAO.

25. Springer to Otero, Jan. 26, 1903, MAO.

26. Typewritten petition, "To the Governor of New Mexico," dated Jan. 28, 1903, MAO.

27. *Record of Proceedings, Las Vegas Normal School,* regents minutes of Jan. 30, 1903, NMHU.

28. *Las Vegas (N.Mex.) Daily Optic,* Jan. 31, 1903.

29. Otero to Springer, Feb. 10, 1903, MAO. Otero's recollection nearly forty years after the fact was that he "had forced Frank Springer to resign as president of the board of regents and appointed Dr. W. R. Tipton in his place." This statement is wrong on two counts and appears to constitute a revision of history. See Miguel Antonio Otero, *My Nine Years as Governor of the Territory of New Mexico, 1897–1906,* 330.

30. Lynn Perrigo, *Gateway to Glorieta,* 143.

31. Kenneth M. Chapman, memoirs, typescript, KMC.

32. *A Statement by the Faculty of the New Mexico Normal University,* 8.

33. Ibid.

34. T. D. A. Cockerell, "The New Mexico Normal University," 746.

35. Hewett, *Two Score Years,* 108.

Chapter 10

1. Murphy, *Philmont,* 147.

2. *Maxwell Land Grant Mining Districts, and the Mining Regulations of the Maxwell Land Grant Company,* FS/CS.

3. Pearson, *The Maxwell Land Grant,* 186.

4. *Memorandum.* A note in Springer's hand indicates Springer as the author and 1893 as the year of issue, FS/CS.

5. Ibid., 23.

6. Pearson, *The Maxwell Land Grant,* 213.
7. Ibid., 214–15.
8. Frank Springer to Wertheim and Gompertz, Aug. 15, 1902, FS/CS.
9. Ibid.
10. Wertheim and Gompertz to Frank Springer, Sept. 2, 1902, FS/CS.
11. "Prospectus. New Mexico & Pacific Railroad Company," 1903.
12. Henry Koehler to Fisk and Robinson, May 10, 1905, FS/CS.
13. *Las Vegas (N.Mex.) Daily Optic,* July 8, 1905.
14. Kenneth Chapman, "Springer Memorial Services at Las Vegas: By Kenneth M. Chapman," 393.
15. Ibid., 392.
16. Twitchell, *Notes and Memoranda,* 14.
17. Frank Springer to H. J. Hagerman, July 31, 1906, Springer Scrapbook, Frank Springer Collection, New Mexico State Records Center and Archives.
18. Springer to Hewett, Dec. 17, 1907, ELH.
19. Twitchell, *Notes and Memoranda,* 14.

Chapter 11

1. Springer, "Address Delivered by Hon. Frank Springer at the Dedication of New Museum Building, Santa Fe, New Mexico; November 25, 1917," *Papers of the School of American Research* 42 (1917), 2.
2. Beatrice Chauvenet, *Hewett and Friends: A Biography of Santa Fe's Vibrant Era,* 209.
3. For an account of Fletcher's life and work, including relations with Hewett and the School of American Archaeology, see Joan Mark, *A Stranger in Her Native Land: Alice Fletcher and the American Indians.*
4. Ibid., 320.
5. Frank Springer to Alice Fletcher, Aug. 25, 1909, FS/CS.
6. Springer to Lummis, June 17, 1912, CFL.
7. *Albuquerque Morning Journal,* Sept. 3, 1915.
8. *Santa Fe New Mexican,* Sept. 6, 1916.
9. J. K. Shishkin, *The Palace of the Governors,* 3–4.
10. W. W. H. Davis, *El Gringo: New Mexico and Her People,* 169; Lew Wallace to Susan Wallace, Oct. 8, 1878, LW; Susan E. Wallace, *The Land of the Pueblos,* 15.
11. See Chauvenet, *Hewett and Friends,* 74–77.
12. Amended House Bill No. 100, "An Act to Establish a Museum for the Territory of New Mexico, and for Other Purposes," *Acts of the Legislative Assembly of the Territory of New Mexico, 1909* (Santa Fe: New Mexican Printing, 1909).
13. *Santa Fe New Mexican,* Apr. 10, 1909.
14. Edgar L. Hewett, "Report of the Director, 1909," in Archaeological Institute of America, *Organic Acts and Administrative Reports of the School of American Archaeology,* 56.
15. *Santa Fe New Mexican,* Feb. 12, Aug. 22, 1910.
16. *Las Vegas (N.Mex.) Optic,* Dec. 10, 1910.
17. "Address of Hon. Frank Springer at Las Vegas, New Mexico, October 7, 1911," privately printed, n.d., FS/CS, reprinted in full in *Santa Fe New Mexican,* Oct. 9, 1911; *Las Vegas (N.Mex.) Optic,* Nov. 4, 1911.
18. See Chauvenet, *Hewett and Friends,* 109–20.

19. *Santa Fe New Mexican,* Oct. 3, 1913.

20. *Santa Fe New Mexican,* Oct. 23, 1913.

21. *Santa Fe New Mexican,* Nov. 12, 1913.

22. "Extract from Minutes of Meeting of the Executive Committee, School of American Archaeology, Held at the National Museum, Washington, D.C., October 30, 1913," FS/CS.

23. J. A. Rolls to Springer, Nov. 8, 1913; Springer to Rolls, Nov. 9, 1913, both in FS/CS.

24. Springer to Rolls, Nov. 8, 1913, FS/CS.

25. Twitchell to Springer, Nov. 10, 1913, FS/CS.

26. *Santa Fe New Mexican,* Nov. 15, 1913.

27. Minutes of Meeting of the Executive Committee, School of American Archaeology, U.S. National Museum, Washington, D.C., Dec. 18, 1913; copy in Frederick Webb Hodge Papers.

Chapter 12

1. Springer, *Uintacrinus: Its Structure and Relations,* Memoirs of the Museum of Comparative Zoology, Harvard College 25:1 (Aug., 1901): 1–89.

2. Charles R. Keyes, "Springer of the Crinoids," *Pan-American Geologist* 48:5 (Dec., 1927): 321–34.

3. Springer to Charles Wachsmuth, Feb. 24, 1890, FS/NMNH.

4. Springer to Alexander Agassiz, Feb. 9, 1902, FS/NMNH.

5. Ibid.

6. Charles Schuchert, "Memorial of Frank Springer," *Bulletin of the Geological Society of America* 39 (1928): 72.

7. In 1908 Stuart Weller of the University of Chicago made a plea for Springer's collection on behalf of the Walker Museum. Stuart Weller to Springer, Nov. 5, 1908, FS/NMNH. Reference is made to consideration given the American Museum of Natural History in a letter from H. C. Bumpus to Springer, May 26, 1909, FS/NMNH.

8. See Ellis L. Yochelson, *The National Museum of Natural History: 75 Years in the Natural History Building,* 9–19.

9. Charles D. Walcott to Springer, Apr. 10, 1907, FS/NMNH.

10. R. S. Bassler to Springer, Nov. 6, 1908, FS/NMNH.

11. Bassler to Springer, June 16, 1911, FS/NMNH.

12. *Burlington (Iowa) Hawk-Eye,* Nov. 12, 1911.

13. Walcott to Springer, Dec. 9, 1911, FS/NMNH.

14. James M. Goode, *Best Addresses: A Century of Washington's Distinguished Apartment Houses,* 95–97.

15. Chapman, "Springer Memorial Services at Las Vegas," 394.

16. "The Crinoidea Flexibilia," review in *El Palacio* 11:4 (Aug. 15, 1921): 51–52.

17. Chapman, "Springer Memorial Services at Las Vegas," 397–98; quotations concerning this incident in the following paragraphs are also from this source.

18. Pearson, *The Maxwell Land Grant,* 238.

19. Minutes of the Annual Meeting of the Board of Trustees, Maxwell Land Grant Company, May 9, 1893, MLG.

20. Minutes of the Annual Meeting of the Board of Trustees, Maxwell Land Grant Company, Apr. 10, 1906, MLG.

21. "Eagle Nest Dam," unsigned typescript, n.d., FS/CS.

22. Frank Springer to Henry M. Porter, Nov. 15, 1918, Henry Miller Porter Papers.

Chapter 13

1. Josephine Springer to Frank Springer, Apr. 4, 1909, FS/CS; Springer to R. S. Bassler, Oct. 18, 1911, RSB.

2. "Santa Fe Society Notes: The Secret of Keeping Young," *Albuquerque Morning Journal,* Oct. 8, 1916.

3. Narrative description and summary of personal assets, dated Oct. 11, 1915, and signed by Springer, FS/CS.

4. Chapman, "Springer Memorial Services at Las Vegas," 396–97.

5. Springer, "Address Delivered by Hon. Frank Springer at the Dedication of New Museum Building," 1917.

6. Hewett to Frank Springer, Apr. 27, 1912, ELH.

7. Springer to Hewett, May 20, 1912; Springer to Hewett, Oct. 14, 1912, both in ELH.

8. Beauregard to Springer, Oct. 29, 1912, ELH.

9. Beauregard to Springer, Dec. 20, 1912; Beauregard to Springer, Sept. 12, 1913, both in ELH.

10. Signed statement by Kenneth Chapman, Apr. 28, 1925, DB.

11. Springer to Kenneth Chapman, March 14, 1914, quoted in "Extracts from Letters of Frank Springer to K. M. Chapman," an unsigned, undated document most likely prepared by Chapman to assist Springer in documenting transactions involving Beauregard's works, DB.

12. Beauregard to Springer, March 27, 1914, ELH; Springer to Kenneth Chapman, Apr. 6, 1914, quoted in "Extracts from Letters of Frank Springer to K. M. Chapman," both in DB.

13. Signed statement by Kenneth Chapman, Apr. 28, 1925, DB.

14. Chris Wilson, *The Myth of Santa Fe: Creating a Modern Regional Tradition.*

15. Springer to Jesse Nusbaum, March 30, 1915, JLN.

16. Handwritten note listing gifts to the museum, including Springer's personal contribution to the building fund, n.d., FS/CS.

17. Chauvenet, *Hewett and Friends,* 124–25.

18. Hewett to Springer, Oct. 14, 1916; Paul A. F. Walter to Springer, Dec. 15, 1916, both in FS/CS.

19. Springer, "Address Delivered by Hon. Frank Springer at the Dedication of New Museum Building," 1917.

20. C. Wilson, *The Myth of Santa Fe,* 70.

21. Edgar L. Hewett, "Santa Fe in 1926," 12.

22. Springer to Walter, Apr. 5, 1918, FS/FACHL.

23. Springer to Walter, Feb. 26, 1918, PAFW.

24. Springer to Walter, Apr. 3, 1918, PAFW.

25. C. Wilson, *The Myth of Santa Fe,* 142.

26. Springer to Hewett, Nov. 11, 1918, ELH.

Chapter 14

1. See Robert R. White, *The Taos Society of Artists.*

2. Frank Springer to Austin H. Clark, July 28, 1914, Austin H. Clark Papers.

3. Springer to Bather, May 24, 1917, NHM.

4. Springer, "Address Delivered by Hon. Frank Springer at the Dedication of New Museum Building," 1917, 1.

5. Springer to Bather, June 5, 1918, NHM.

6. Springer to "My Dear Family," July 11, 1918, FS/CS.

7. Springer to Daniel T. Hoskins, July 11, 1918, FS/CS.

8. Springer to "My Dear Family," July 11, 1918, FS/CS.

9. Springer to Bather, Apr. 17, 1920, NHM.

10. "In Memoriam: Henry Strong Springer," signed "His Parents," brochure, privately printed, Feb. 12, 1920, FS/CS.

11. Springer to Jesse Nusbaum, Feb. 22, 1920, JLN.

12. An account of this matter is found in Price, "A History of the New Mexico Normal University"; unless otherwise noted all quotations concerning the groundbreaking ceremony are from Price.

13. Ibid., 81.

14. Draft minutes of the regents meeting of Oct. 9, 1919, sent to Springer as an enclosure with letter from Frank H. H. Roberts to Frank Springer, Oct. 18, 1919, FS/CS.

15. Frank Springer to Eva Springer, Apr. 19, 1924, FS/CS.

16. "Presentation of Bronze Bust of Hon. Frank Springer," *El Palacio* 13:6–7 (October 2, 1922): 75–88; the quote from Hewett is on p. 86.

17. Wallace Springer to Ada Springer Davis, Sept. 19, 1922, FS/CSWR.

18. Edward T. Springer to C. F. Lummis, Oct. 7, 1923; Lummis to Edward T. Springer, Oct. 10, 1923, both in CFL.

19. Frank Springer to Charles Schuchert, Nov. 11, 1923, CSP.

20. Springer to Lummis, Jan. 23, 1924, CFL.

21. Regents of the Museum of New Mexico, Board of Managers, School of American Research, Minutes of Sept. 4, 1924, Official Minute Books, Administrative Offices, Museum of New Mexico, Santa Fe.

22. Springer to Schuchert, Oct. 2, 1924, CSP.

Chapter 15

1. Springer to William H. King, Apr. 30, 1925, DB.

2. Springer to E. L. Hewett, July 14, 1925, DB.

3. See Don D. Fowler, *A Laboratory for Anthropology: Science and Romanticism in the American Southwest, 1846–1930,* for a detailed account of this episode.

4. Kenneth Chapman to Springer, Sept. 1, 1926; Chapman's handwritten copy of the letter was retained for his files. KMC. Springer to Chapman, Nov. 22, 1926, KMC.

5. Information on the origins of the work is found in Springer's introduction to *American Silurian Crinoids,* 1926.

6. Springer to F. A. Bather, Apr. 20, 1924, NHM.

7. Springer to Ray S. Bassler, Dec. 6, 1924, RSB.

8. Springer to Paul A. F. Walter, Dec. 9, 1926, PAFW.

9. Charles Schuchert, review of *American Silurian Crinoids,* by Frank Springer, *Science* 65:1687 (Apr. 29, 1927): 425–26.

10. Springer to Jan van Houten, Dec. 4, 1926, FS/CS.

11. H. A. Hartogh to Springer, Jan. 20, 1927, FS/CS.

12. van Houten to Springer, Aug. 8, 1927, FS/CS.

13. In response to information requested by the Bradstreet Company, Arthur H. Officer had written on Oct. 4, 1926, advising, "The property of the corporation now held in New Mexico is valued at about $800,000, consisting of lands, timber, mineral and mining. The corporation has been in a process of liquidation for a number of years. It has no outstanding current liabilities, the exception being its bonded debt held abroad," MLG.

14. After Springer's death, his daughter Ada wrote to the Hewetts, giving an account of Springer's final decline. Ada Springer Davis to Dr. and Mrs. E. L. Hewett, Oct. 5, 1927, ELH.

15. Frank Springer to Edward Springer, Aug. 3, 1927, FS/CS.

16. Ada Springer Davis to Dr. and Mrs. E. L. Hewett, Oct. 5, 1927, ELH.

17. Ibid.

18. Reid, "The Springer Memorial Services," 371.

19. Orin L. Phillips, "The Springer Memorial Services: Eulogy by Hon. Orin L. Phillips," *El Palacio* 23:14–16 (Oct. 22, 1927): 377.

20. Chapman to Josephine Springer, Sept. 25, 1927, FS/CS.

Chapter 16

1. Minutes of the Board of Trustees, Maxwell Land Grant Company, Oct. 15, 1927, MLG.

2. Information concerning Springer's estate is from a copy of the "Return for Federal Estate Tax" prepared after Springer's death, Dec. 8, 1927, FS/CS.

3. *Raton (N.Mex.) Range,* July 26, 1955.

4. See James E. Snead, *Ruins and Rivals: The Making of Southwest Archaeology.*

5. Raymond C. Moore, dedication, "Part T: Echinodermata," in *Treatise on Invertebrate Paleontology,* ed. Moore and Curt Teichert, 1:73; William I. Ausich, "Lower Mississippian Burlington Limestone along the Mississippi River Valley in Iowa, Illinois, and Missouri, USA," in Hans Hess et al., *Fossil Crinoids,* 139.

6. Ada Springer Davis to Charles Schuchert, n.d., CSP.

7. Frank Springer, "Address Delivered by Hon. Frank Springer at the Dedication of New Museum Building," 1917.

Selected Works of Frank Springer

Unless otherwise noted, Frank Springer is the sole author of works in this bibliography.

1877

Wachsmuth, Charles, and Frank Springer. "Revision of the Genus *Belemnocrinus,* and Description of Two New Species." *American Journal of Science* 3rd ser., 13:76 (Apr.): 253–60.

1879

Wachsmuth, Charles, and Frank Springer. *Revision of the Palaeocrinoidea.* Pt. 1. *Proceedings of the Academy of Natural Sciences of Philadelphia,* separate printing.

———. "Transition Forms in Crinoids and Description of Five New Species." *Proceedings of the Academy of Natural Sciences of Philadelphia, 1878,* 224–66.

1881

Wachsmuth, Charles, and Frank Springer. *Revision of the Palaeocrinoidea.* Pt. 2. *Proceedings of the Academy of Natural Sciences of Philadelphia,* separate printing.

1883

Wachsmuth, Charles, and Frank Springer. "On *Hybocrinus, Hoplocrinus,* and *Baerocrinus.*" *American Journal of Science* 3rd ser., 26:155 (Nov.): 365–77.

———. "Remarks on *Glyptocrinus* and *Reteocrinus,* Two Genera of Silurian Crinoids." *American Journal of Science* 3rd ser., 25:148 (Apr.): 255–68.

1884

"Occurrence of the Lower Burlington Limestone in New Mexico." *American Journal of Science* 3rd ser., 27:158 (Feb.): 97–103.

1885

Wachsmuth, Charles, and Frank Springer. *Revision of the Palaeocrinoidea.* Pt. 3. sec. 1. *Proceedings of the Academy of Natural Sciences of Philadelphia,* separate printing.

1886

Wachsmuth, Charles, and Frank Springer. *Revision of the Palaeocrinoidea.* Pt. 3, sec. 2. *Proceedings of the Academy of Natural Sciences of Philadelphia,* separate printing.

1887

Conclusion of Oral Argument in *United States v. Maxwell Land-Grant Co. et al,* 121 U.S. 325. In *The Spanish Archives of New Mexico,* 1:53–60. Cedar Rapids, Iowa: Torch Press, 1914.

Springer, Frank, and Charles E. Gast. *Brief for Appellees, United States v. Maxwell Land-Grant Company et al.* Cause No. 974, Supreme Court of the United States, October Term, 1886. Washington, D.C., Gibson Bros.

1888

Wachsmuth, Charles, and Frank Springer. "The Summit Plates in Blastoids, Crinoids, and Cystids, and Their Morphological Relations." *Proceedings of the Academy of Natural Sciences of Philadelphia, 1887,* 82–114.

1889

Wachsmuth, Charles, and Frank Springer. "*Crotalocrinus:* Its Structure and Zoological Position." *Proceedings of the Academy of Natural Sciences of Philadelphia, 1888,* 364–90.

———. "Discovery of the Ventral Structure of *Taxocrinus* and *Haplocrinus,* and Consequent Modification in the Classification of the Crinoidea." *Proceedings of the Academy of Natural Sciences of Philadelphia, 1888,* 337–63.

1890

"Land Titles in New Mexico." Annual address of the retiring president of the New Mexico Bar Association, Jan. 7, 1890. Santa Fe: New Mexican Printing.

Wachsmuth, Charles, and Frank Springer. "New Species of Crinoids and Blastoids from the Kinderhook Group of Le Grand, Iowa; and A New Genus from the Niagara Group of Western Tennessee." *Illinois Geological Survey* 8 (July): 155–208.

1891

Wachsmuth, Charles, and Frank Springer. "The Perisomic Plates of the Crinoids." *Proceedings of the Academy of Natural Sciences of Philadelphia, 1890,* 345–92.

1892

Wachsmuth, Charles, and Frank Springer. "Description of Two New Genera and Eight Species of Camerate Crinoids from the Niagara Group." *American Geologist* 10:3 (Sept.): 135–44.

———. *The North American Crinoidea Camerata.* Cambridge: Memoirs of the Museum of Comparative Zoology, Harvard College 20 and 21 (May).

1899

"Dedicatory Address, New Mexico Normal University at Las Vegas." *Southwest* 1:4 (March): 84–89.

"Presentation Address." Address to the first graduating class of New Mexico Normal University. *Southwest* 1:7 (June): 167–70.

1900

"On the Presence of Pores in the Ventral Sac in Fistulate Crinoids." *American Geologist* 26:3 (Sept.): 133–51.

1901

"John Marshall Day Oration." *Las Vegas (N.Mex.) Daily Optic* (Feb. 5).

"The Olympic Games." Address to graduates of New Mexico Normal University. *Las Vegas (N.Mex.) Daily Optic* (June 21).

Uintacrinus: Its Structure and Relations. Cambridge: Memoirs of the Museum of Comparative Zoology, Harvard College 25:1 (Aug.).

1902

"On the Crinoid Genera *Sagenocrinus, Forbesiocrinus,* and Allied Forms." *American Geologist* 30:2 (Aug.): 88–97.

1904

"*Actinometra iowensis,* a New Unstalked Crinoid from the Florida Reefs." *Bulletin, Laboratories of Natural History, State University of Iowa* 5:4, 219–23.

1905

Cleocrinus. Cambridge: Memoirs of the Museum of Comparative Zoology, Harvard College 25:2 (Jan.): 93–112.

1906

"Discovery of the Disk of *Onychocrinus,* and Further Remarks on the Crinoidea Flexibilia." *Journal of Geology* 14:6 (Sept.–Oct.): 467–523.

Springer, Frank, and Arthur Ware Slocum. "*Hypsocrinus,* a New Genus of Crinoids from the Devonian." Chicago: Field Columbian Museum, Publication 114, Geological Series 2:9, 267–71.

1909

"A New American Jurassic Crinoid." Publication 1664. *Proceedings of the U.S. National Museum* 36 (March 3): 179–90.

1910

"The Field Session of the School of American Archaeology." *Science* 32:827 (Nov. 4): 622–24.

1911

"Address of Hon. Frank Springer at Las Vegas, New Mexico, October 7, 1911." *Santa Fe New Mexican* (Oct. 9).

"The Crinoid Fauna of the Knobstone Formation." Publication 1850. *Proceedings of the U.S. National Museum* 41 (June 24): 175–208.

On a Trenton Echinoderm Fauna at Kirkfield, Ontario. Ottawa: Memoirs of the Geological Survey of Canada 15-P.

Some New American Fossil Crinoids. Cambridge: Memoirs of the Museum of Comparative Zoology, Harvard College 25:3 (July).

1913

"*Pelmatozoa.*" Revised section including discussion of the Cystoidea, Blastoidea, and Crinoidea. In *Text-book of Paleontology,* ed. Karl A. Von Zittel and Charles R. Eastman. 2nd ed. rev. London: Macmillan.

1917

"Address Delivered by Hon. Frank Springer at Dedication of New Museum Building, Santa Fe, N.M., November 25, 1917." *Papers of the School of American Research* 42.

On the Crinoid Genus Scyphocrinus *and Its Bulbous Root,* Camarocrinus. Publication 2440. Washington, D.C.: Smithsonian Institution.

1918

"A New Species of Fossil *Pentacrinus* from the East Indies." Nederlandische Timor-Expeditie 2. *Jaarboek van het Mijnwesen* 45 (Jan.): 1–8.

"On *Mystiocrinus,* a New Genus of Silurian Crinoidea." *American Journal of Science* 4th ser., 46:275 (Nov.): 666–69.

1919

"New Species of Crinoid: *Eucheirocrinus Ontario.*" In *The Silurian Geology and Faunas of Ontario Peninsula, and Manitoulin and Adjacent Islands,* by M. Y. Williams. Geological Survey of Canada, Memoir 111 (Geological Series 91): 127.

"Oration by Frank Springer, Theodore Roosevelt Memorial Service." *Papers of the School of American Research* 43. Also published as a complete issue of *El Palacio* 6:5 (Feb. 22, 1919).

1920

The Crinoidea Flexibilia. 2 vols. Publication 2501. Washington, D.C.: Smithsonian Institution.

1921

The Fossil Crinoid Genus Dolatocrinus *and Its Allies*. Bulletin 115. Washington, D.C.: U.S. National Museum.

"New Species of Devonian Crinoidea from Northern Canada." Bulletin No. 33. Geological Survey of Canada, Victoria Memorial Museum.

1922

"Crinoids from the Upper Cretaceous of Tamaulipas, Mexico." Publication 2426. *Proceedings of the U.S. National Museum* 61 (Apr. 25): 1–4.

1923

On the Fossil Crinoid Family Catillocrinidae. Publication 2718. Washington, D.C.: Smithsonian Miscellaneous Collections 76:3 (Aug. 3).

1924

"The Genus *Pentacrinus* in Alaska." Publication 2577. *Proceedings of the U.S. National Museum* 67, 1–7.

"On the Genus *Holopus*." *University of Iowa Studies in Natural History* (Aug.): 45–63.

"A Tertiary Crinoid from the West Indies." Publication 2516. *Proceedings of the U.S. National Museum* 65 (May 12): 1–8.

1925

"Occurrence of the Crinoid Genus *Apiocrinus* in America." Publication 2590. *Proceedings of the U.S. National Museum* 67, 1–5.

1926

American Silurian Crinoids. Publication 2871. Washington, D.C.: Smithsonian Institution.

"Unusual Forms of Fossil Crinoids." Publication 2581. *Proceedings of the U.S. National Museum* 67, 1–137.

"Upper Devonian Crinoids from Mackenzie River Valley." Bulletin No. 42. Contributions to Canadian Paleontology, Geological Survey of Canada, 127–32.

1927

Educational Addresses of Dr. Frank Springer. Papers of the School of American Research. N.s., paper no. 17 (Oct.).

Bibliography

The bibliography is divided into the following sections: "Primary Sources," "Secondary Sources," "Newspapers," "Records of Legal Proceedings." A separate bibliography lists selected works authored or coauthored by Frank Springer.

Primary Sources

Agassiz, Alexander. Papers. Museum of Comparative Zoology Archives, Ernst Mayr Library, Harvard University, Cambridge, Mass.

Bassler, Ray S. Papers. Smithsonian Institution Archives, Washington, D.C.

Beauregard, Donald. Papers. Fray Angélico Chávez History Library, Palace of the Governors, Museum of New Mexico, Santa Fe.

Bowditch, Charles P. Papers. Peabody Museum of Archaeology and Ethnology, Harvard University, Cambridge, Mass.

Catron, Thomas B. Papers. Center for Southwest Research, University of New Mexico, Albuquerque.

Chapman, Kenneth M. Papers. Indian Arts Research Center, School of American Research, Santa Fe.

Clark, Austin H. Papers. Smithsonian Institution Archives, Washington, D.C.

Fletcher, Alice Cunningham. Papers. National Anthropological Archives, Smithsonian Institution, Washington, D.C.

Hewett, Edgar Lee. Papers. Fray Angélico Chávez History Library, Palace of the Governors, Museum of New Mexico, Santa Fe.

Hodge, Frederick Webb. Papers. Braun Research Library, Autry National Center/Southwest Museum, Los Angeles.

Jones, Andrieus A. Papers. New Mexico State Records Center and Archives, Santa Fe.

Lummis, Charles Fletcher. Papers. Autry National Center/Southwest Museum, Los Angeles.

Maxwell Land Grant Collection. Center for Southwest Research, University of New Mexico, Albuquerque.

National Archives and Records Administration, Washington, D.C. RG 48.

Natural History Museum, London, UK. Department of Palaeontology, Departmental Correspondence, Correspondence between Francis Arthur Bather and Frank Springer, DF100/203.

New Mexico Highlands University. University Archives, Las Vegas, N.Mex.

Nusbaum, Jesse Logan. Papers. National Anthropological Archives, Smithsonian Institution, Washington, D.C.

Official Minute Books. Administrative Offices. Museum of New Mexico, Santa Fe.

Otero, Miguel A. Papers. New Mexico State Records Center and Archives, Santa Fe.

Porter, Henry Miller. Papers. Colorado Historical Society, Denver, Colo.

Records of the Surveyor General of the Territory of New Mexico. United States General Land Office, National Archives and Records Administration, Surveyor General 15.

Rockefeller Family Collection. Rockefeller Archive Center, Sleepy Hollow, N.Y. Kenneth Chapman, "My First Quarter-Century in the Southwest," typescript, March 27, 1954.

Schuchert, Charles. Papers. Yale University Library, New Haven, Conn.

Springer, Frank. Collection. New Mexico State Records Center and Archives, Santa Fe.

Springer, Frank. Papers. Center for Southwest Research, University of New Mexico, Albuquerque.

Springer, Frank. Papers. CS Cattle Company, Cimarron, N.Mex.

Springer, Frank. Papers. Fray Angélico Chávez History Library, Palace of the Governors, Museum of New Mexico, Santa Fe.

Springer, Frank. Papers. National Museum of Natural History, Department of Paleobiology, Smithsonian Institution, Washington, D.C.

Territorial Archives of New Mexico. New Mexico State Records Center and Archives, Santa Fe.

Wallace, Lew. Collection. Indiana Historical Society, Indianapolis.

Walter, Paul A. F. Papers. Fray Angélico Chávez History Library, Palace of the Governors, Museum of New Mexico, Santa Fe.

Secondary Sources

Acts of the Legislative Assembly of the Territory of New Mexico, 1909. Santa Fe: New Mexican Printing, 1909.

Anderson, Wayne I. *Iowa's Geological Past: Three Billion Years of Earth History*. Iowa City: University of Iowa Press, 1998.

Anderson, Wayne I., and William M. Furnish. "Iowa's Self-Trained Paleontologists." *Proceedings of the Iowa Academy of Science* 90:1 (1983): 1–12.

Archaeological Institute of America. *Organic Acts and Administrative Reports of the School of American Archaeology, 1907–1917*. Santa Fe, 1917.

Armstrong, Ruth W. *The Chases of Cimarron: Birth of the Cattle Industry in Cimarron Country, 1867–1900*. Albuquerque: New Mexico Stockman, 1981.

Ausich, William I., and Thomas W. Kammer. "The Study of Crinoids during the 20th Century and the Challenges of the 21st Century." *Journal of Paleontology* 75:6 (2001): 1161–73.

Bassler, R. S., and Margaret W. Moodey. *Bibliographic and Faunal Index of Paleozoic Pelmatozoan Echinoderms*. Geological Society of America Special Papers, No. 45. New York: Geological Society of America, 1943.

Bloom, Lansing B. "Charles Springer." *Dictionary of American Biography*. Vol 11. New York: Charles Scribner's Sons, 1928.

Bradfute, Richard Wells. *The Court of Private Land Claims: The Adjudication of Spanish and Mexican Land Grant Titles, 1891–1904*. Albuquerque: University of New Mexico Press, 1975.

Bryant, Keith L., Jr. *History of the Atchison, Topeka, and Santa Fe Railway*. New York: Macmillan, 1974.

Calvin, Samuel. "Memoir of Charles Wachsmuth." *Bulletin of the Geological Society of America* 8 (Apr. 30, 1897): 374–76.

Caperton, Thomas J. *Rogue: Being an Account of the Life and High Times of Stephen W. Dorsey, United States Senator and New Mexico Cattle Baron*. Santa Fe: Museum of New Mexico Press, 1978.

Chapman, K. M. "The Cave Pictographs of the Rito de Los Frijoles, N.M." and "Conservatism in the Evolution of Decorative Art." *Papers of the School of American Archaeology* No. 37 (1917).

———. "Springer Memorial Services at Las Vegas: By Kenneth M. Chapman." *El Palacio* 23: 14–16 (Oct. 22, 1927): 389–99.

Chase, C. M. *The Editor's Run in New Mexico and Colorado*. Montpelier: Argus and Patriot Book and Job Printing House, 1882.

Chauvenet, Beatrice. *Hewett and Friends: A Biography of Santa Fe's Vibrant Era*. Santa Fe: Museum of New Mexico Press, 1983.

Cleaveland, Agnes Morley. *No Life for a Lady*. Boston: Houghton Mifflin, 1941.

Cleaveland, Norman. *An Introduction to the Colfax County War*. Privately printed, n.d. Copy in collection of Center for Southwest Research, University of New Mexico, Albuquerque.

Cleaveland, Norman, with George Fitzpatrick. *The Morleys: Young Upstarts on the Southwest Frontier*. Albuquerque: Calvin Horn, 1971.

Cockerell, T. D. A. "The New Mexico Academy of Science." *Science* n.s. 17:420 (1903): 114.

———. "The New Mexico Normal University." *Science* n.s. 17:436 (1903): 746–47.

"The Crinoidea Flexibilia." Review in *El Palacio* 11:4 (Aug. 15, 1921): 49–52.

Davis, W. W. H. *El Gringo: New Mexico and Her People*. New York: Harper and Brothers, 1857.

DeVolder, Arthur L. "John Gaw Meem, F.A.I.A.: An Appreciation." *New Mexico Historical Review* 54:3 (July, 1979): 209–25.

Dunham, Harold H. *Government Handout: A Study in the Administration of Public Lands, 1875–1891*. Ann Arbor: Edwards Brothers, 1941.

Ebright, Malcolm. *Land Grants and Lawsuits in Northern New Mexico*. Albuquerque: University of New Mexico Press, 1994.

Elliott, Melinda. *Exploring Human Worlds: A History of the School of American Research*. Santa Fe: School of American Research, 1987.

Fergusson, Harvey. *Grant of Kingdom*. New York: William Morrow, 1950.

Foster, Mark S. *Henry M. Porter: Rocky Mountain Empire Builder*. Niwot: University Press of Colorado, 1991.

Fowler, Don D. *A Laboratory for Anthropology: Science and Romanticism in the American Southwest, 1846–1930*. Albuquerque: University of New Mexico Press, 2000.

Gerber, John C., Carolyn B. Brown, James Kaufman, and James B. Lindberg Jr. *A Pictorial History of the University of Iowa*. Iowa City: University of Iowa Press, 1988.

Goode, James M. *Best Addresses: A Century of Washington's Distinguished Apartment Houses*. Washington, D.C.: Smithsonian Institution Press, 1988.

Hammond, William Gardiner. *Records, Iowa School of Law, 1866–1878*. Handwritten journal in the Rare Book Collection, Law Library, College of Law, University of Iowa, Iowa City.

Hansen, Millard Winchester. "The Early History of the College of Law, State University of Iowa: 1865–1884." *Iowa Law Review* 30:1 (Nov., 1944): 31–67.

Harwood, Thomas. *History of New Mexico Spanish and English Missions of the Methodist Episcopal Church from 1850 to 1910*. Vol. 1. Albuquerque: El Abogado Press, 1908.

Hess, Hans, William I. Ausich, Carlton E. Brett, and Michael J. Simms. *Fossil Crinoids*. Cambridge and New York: Cambridge University Press, 1999.

Hewett, Edgar L. *Baccalaureate Addresses (1899–1903)*. San Diego: Arts and Crafts Press, 1903.

———. "Santa Fe in 1926." *Papers of the School of American Archaeology* 39 (1917): 8–12.

———. "The Springer Memorial Services: Address of Dr. Edgar L. Hewett." *El Palacio* 23:14–16 (Oct. 22, 1927): 378–88.

———. *Two Score Years*. Albuquerque: University of New Mexico Press, 1946.

Holmes, Frank. "The Horses of the CS Ranch." Pt. 1. *Western Horseman* 60:11 (Nov., 1995): 161–68.

———. "The Horses of the CS Ranch." Pt. 2. *Western Horseman* 60:12 (Dec., 1995): 14–20.

"In Memoriam: Henry Strong Springer." Brochure, privately printed, attributed to "His Parents." Feb. 12, 1920. Frank Springer Papers, CS Cattle Company, Cimarron, N.Mex.

"In Memoriam: Lt. Lew Wallace Springer." *El Palacio* 24:13–14 (March 31–Apr. 7, 1928): 245–47.

Judd, Neil M. "1910 in El Rito de los Frijoles." *El Palacio* 69:3 (Fall, 1962): 139–41.

Julian, George W. "Land Stealing in New Mexico." *North American Review* 145:368 (1887): 17–31.

Keleher, William A. *Maxwell Land Grant: A New Mexico Item.* Santa Fe: Rydal Press, 1942.

———. *Violence in Lincoln County, 1869–1881: A New Mexico Item.* Albuquerque: University of New Mexico Press, 1957.

Keyes, Charles Rollin. "An Epoch in the History of American Science." *Annals of Iowa* 3rd ser., 2 (Apr., 1896): 345–64.

———. Review of *North American Fossil Crinoidea Camerata,* by Charales Wachsmuth and Frank Springer. *Journal of Geology* 4:2 (Feb.–March, 1896): 221–40.

———. "Springer of the Crinoids." *Pan-American Geologist* 48:5 (Dec., 1927): 321–34.

Keyes, Margaret N. *Old Capitol: Portrait of an Iowa Landmark.* Iowa City: University of Iowa Press, 1988.

Lamar, Howard Roberts. "Edmund G. Ross as Governor of New Mexico Territory: A Reappraisal." *New Mexico Historical Review* 36:3 (July, 1961): 196–97.

———. *The Far Southwest, 1846–1912: A Territorial History.* New Haven: Yale University Press, 1966.

Lambert, Oscar Doane. *Stephen Benton Elkins: American Foursquare.* Pittsburgh: University of Pittsburgh Press, 1955.

Larson, Robert W. *New Mexico's Quest for Statehood, 1846–1912.* Albuquerque: University of New Mexico Press, 1968.

Lee, Ronald Freeman. "The Antiquities Act of 1906." Ed. Raymond Harris Thompson. *Journal of the Southwest* 42:2 (Summer, 2000): 197–270.

The Legislative Blue Book of the Territory of New Mexico. Santa Fe: Charles W. Greene, Public Printer, 1882.

Leopold, Aldo. *A Sand County Almanac, with Other Essays on Conservation from Round River.* Enlarged ed. New York: Oxford University Press, 1966.

Limerick, Patricia Nelson. *Something in the Soil: Legacies and Reckonings in the New West.* New York: W. W. Norton, 2000.

Making Your Beef: CS. Cimarron: Charles Springer Cattle Co., n.d. Frank Springer Papers, CS Cattle Company, Cimarron, N.Mex.

Mark, Joan. *A Stranger in Her Native Land: Alice Fletcher and the American Indians.* Lincoln: University of Nebraska Press, 1988.

Marshall, James. *Santa Fe: The Railroad That Built an Empire.* New York: Random House, 1941.

Mattson, Hans. *Reminiscences: The Story of an Emigrant.* Saint Paul: D. D. Merrill, 1891.

Maxwell Land Grant Mining Districts and the Mining Regulations of the Maxwell Land Grant Company. Chicago: Printed for the Maxwell Land Grant Company by Rand, McNally, May, 1897. Frank Springer Papers, CS Cattle Company, Cimarron, N.Mex.

"The Maxwell Land Grant: Opinion of the Attorney General of the United States, Declining to Bring Suit to Vacate the Patent, November 19, 1881." Privately printed for Frank Springer, n.d. Frank Springer Papers, CS Cattle Company, Cimarron, N.Mex.

McMains, O. P. *Conspiracy to Defraud.* Broadside dated Apr. 24, 1886. Privately printed. Frank Springer Papers, CS Cattle Company, Cimarron, N.Mex.

———. *Defiance: And Contempt for That Which Is Contemptible.* Broadside c. Oct., 1888. Privately printed. Pamphlet and Tract Collection, University of Colorado Library.

McNary, James Graham. *This Is My Life.* Albuquerque: University of New Mexico Press, 1956.

Memorandum. A twenty-three page printed document making a case for heavy investment in coal resources of the Maxwell Land Grant. Frank Springer is the likely author and 1893 the likely date of issue. Frank Springer Papers, CS Cattle Company, Cimarron, N.Mex.

Montoya, María E. *Translating Property: The Maxwell Land Grant and the Conflict over Land in the American West, 1840–1900.* Berkeley: University of California Press, 2002.

Moore, Raymond C., and Curt Teichert, eds. *Treatise on Invertebrate Paleontology.* Boulder and Lawrence: Geological Society of America and University of Kansas, 1978.

Murphy, Lawrence R. *Lucien Bonaparte Maxwell: Napoleon of the Southwest.* Norman: University of Oklahoma Press, 1983.

———. *Philmont: A History of New Mexico's Cimarron Country.* Albuquerque: University of New Mexico Press, 1972.

Myrick, David F. *New Mexico's Railroads: An Historical Survey.* Golden: Colorado Railroad Museum, 1970.

"New Building, Auditorium, and Art Gallery of the Museum of New Mexico and School of American Archaeology." Pamphlet issued to commemorate the announcement of plans to build the Museum of Fine Arts of the Museum of New Mexico. Frank Springer Papers, CS Cattle Company, Cimarron, N.Mex.

"The New Museum of Santa Fé." Promotional pamphlet for the Museum of Fine Arts of the Museum of New Mexico. n.d. Frank Springer Papers, CS Cattle Company, Cimarron, N.Mex.

Nieuwenhuisen, J. T. "To the Readers of the *News and Press*." Reprint of a letter to the *Cimarron (N.Mex.) News and Press,* originally dated July 20, 1881. Utrecht, n.d. Printed by Kemink and Son.

Otero, Miguel Antonio. *My Life on the Frontier, 1864–1882: Incidents and Characters of the Period When Kansas, Colorado, and New Mexico Were Passing through the Last of Their Wild and Romantic Years.* New York: Press of the Pioneers, 1935.

———. *My Nine Years as Governor of the Territory of New Mexico, 1897–1906.* Albuquerque: University of New Mexico Press, 1940.

Pearson, Jim Berry. *The Maxwell Land Grant.* Norman: University of Oklahoma Press, 1961.

Perrigo, Lynn, *Gateway to Glorieta.* Boulder: Pruett, 1982.

Peterson, William J. *The Story of Iowa: The Progress of an American State.* Vol. 1. New York: Lewis Historical Publishing, 1952.

Phillips, Orin L. "The Springer Memorial Services: Eulogy by the Hon. Orin L. Phillips." *El Palacio* 23:14–16 (Oct. 22, 1927): 374–78.

Poldervaart, Arie W. *Black-Robed Justice.* Santa Fe: Historical Society of New Mexico, 1948.

Porter, Dean, et al. *Taos Artists and Their Patrons, 1898–1950.* South Bend: Snite Museum of Art, University of Notre Dame, 1999.

Porter, Henry M. *Pencilings of an Early Western Pioneer.* Denver: World Press, 1929.

"Presentation of Bronze Bust of Hon. Frank Springer." *El Palacio* 13:6–7 (Oct. 2, 1922): 75–88.

Price, Hugh W. "A History of the New Mexico Normal University, 1893–1931." Master's thesis, New Mexico Normal University, 1932. Special Collections, New Mexico Highlands University General Library, Las Vegas, N.Mex.

"Prospectus, New Mexico and Pacific Railroad Company." Privately printed, 1903. Frank Springer Papers, CS Cattle Company, Cimarron, N.Mex.

Reid, W. C. "The Springer Memorial Services: Address by Captain W. C. Reid." *El Palacio* 23:14–16 (Oct. 22, 1927): 370–74.

Review of *American Silurian Crinoids. El Palacio* 22:1 (Jan. 1, 1927): 2–7.

Sage, Leland L. *A History of Iowa.* Ames: Iowa State University Press, 1974.

Salter, William. *Iowa: The First Free State in the Louisiana Purchase.* Chicago: A. C. McClurg, 1905.

Schuchert, Charles. "Memorial of Frank Springer." *Bulletin of the Geological Society of America* 39 (1928): 65–80.

———. Review of *American Silurian Crinoids,* by Frank Springer. *Science* 65:1687 (Apr. 29, 1927): 425–26.

Sheppard, Carl D. *Creator of the Santa Fe Style: Isaac Hamilton Rapp, Architect.* Albuquerque: University of New Mexico Press, 1988.

———. *The Saint Francis Murals of Santa Fe: The Commission and the Artists.* Santa Fe: Sunstone Press, 1989.

Shishkin, J. K. *The Palace of the Governors.* Santa Fe: Museum of New Mexico, 1972.

Smith, Toby. *Coal Town: The Life and Times of Dawson, New Mexico.* Santa Fe: Ancient City Press, 1993.

Snead, James E. *Ruins and Rivals: The Making of Southwest Archaeology.* Tucson: University of Arizona Press, 2001.

Springer, Francis. "The Address of Hon. Francis Springer." In "A Historic Reunion," *Annals of Iowa* 3rd ser., 3 (Apr., 1897): 36–46.

———. "Recollections of Judge Francis Springer." *Annals of Iowa* 3rd ser., 2 (Jan., 1897): 568–85.

Stanley, F. [Stanley Francis Louis Crocchiola]. *The Grant That Maxwell Bought.* Denver: World Press, 1952.

———. *The Las Vegas, New Mexico, Story.* Denver: World Press, 1951.

A Statement by the Faculty of the New Mexico Normal University, February 1903. Las Vegas, N.Mex.: Optic Print, 1903. Fray Angélico Chávez History Library, Palace of the Governors, Museum of New Mexico, Santa Fe.

Stiles, Edward H. *Recollections and Sketches of Notable Lawyers and Public Men of Iowa; Belonging to the First and Second Generations, with Anecdotes and Incidents Illustrative of the Times.* Des Moines: Homestead, 1916.

Stratton, Porter A. *The Territorial Press of New Mexico, 1834–1912.* Albuquerque: University of New Mexico Press, 1969.

Taylor, Morris F. *First Mail West: Stagecoach Lines on the Santa Fe Trail.* Albuquerque: University of New Mexico Press, 1971.

———. "The Maxwell Cattle Company, 1881–1888." *New Mexico Historical Review* 49:4 (Fall, 1974): 289–324.

———. *O. P. McMains and the Maxwell Land Grant Conflict.* Tucson: University of Arizona Press, 1979.

Thompson, Raymond Harris. "Edgar Lee Hewett and the Political Process." *Journal of the Southwest* 42:2 (Summer, 2000): 271–318.

Tiller, Veronica E. Velarde. *The Jicarilla Apache Tribe: A History.* Rev. ed. Lincoln: University of Nebraska Press, 1992.

Turner, David G., et al. *Artists of 20th-Century New Mexico: The Museum of Fine Arts Collection.* Santa Fe: Museum of New Mexico Press, 1992.

Twitchell, Ralph Emerson. *The Leading Facts of New Mexican History*. 5 vols. Cedar Rapids, Iowa: Torch Press, 1911–17.

———. *Notes and Memoranda Concerning the Legal, Literary, and Scientific Activities of Mr. Frank Springer as a Citizen of New Mexico*. Privately printed for presentation on the occasion of the dedication of the Scarpitta bust of Springer, Museum of New Mexico, 1922. Frank Springer Papers, CS Cattle Company, Cimarron, N.Mex.

Van Sant, Jan F., and N. Gary Lane. *Echinodermata*. University of Kansas Paleontological Contributions, Article 7. Lawrence: University of Kansas Publications, 1964.

Vierra, Carlos. "New Mexico Architecture." *Art and Archaeology* 7:1–2 (Jan.–Feb., 1918): 37–49.

———. "Our Native Architecture in Its Relation to Santa Fe." *Papers of the School of American Archaeology* 39 (1917): 1–7.

Wallace, Susan E. *The Land of the Pueblos*. New York: John B. Alden, 1889.

Waters, L. L. *Steel Trails to Santa Fe*. Lawrence: University of Kansas Press, 1950.

Weber, William A. *The American Cockerell: A Naturalist's Life, 1866–1948*. Boulder: University Press of Colorado, 2000.

Webster, Gary D. *Bibliography and Index of Paleozoic Crinoids, 1942–1968*. Boulder: Geological Society of America, 1973.

Westphall, Victor. *Mercedes Reales: Hispanic Land Grants of the Upper Rio Grande Region*. Albuquerque: University of New Mexico Press, 1983.

———. *Thomas Benton Catron and His Era*. Tucson: University of Arizona Press, 1973.

White, Robert R., ed. *The Taos Society of Artists*. Albuquerque: University of New Mexico Press, 1983.

Wilson, Chris. *The Myth of Santa Fe: Creating a Modern Regional Tradition*. Albuquerque: University of New Mexico Press, 1997.

Wilson, Delphine Dawson. *John Barkley Dawson: Pioneer, Cattleman, Rancher*. Privately printed, 1997.

Winship, A. E. "New Mexico Annual Meeting." *New England Journal of Education* 57:4 (Jan. 22, 1903): 56–57.

"Work Begun on New Museum Building." *El Palacio* (Apr., 1916): 58–61.

Yochelson, Ellis. *The National Museum of Natural History: 75 Years in the Natural History Building*. Washington, D.C.: Smithsonian Institution Press, 1985.

Newspapers

Albuquerque Morning Journal, 1916.

Burlington (Iowa) Daily Gazette, 1871–96.

Burlington (Iowa) Hawk-Eye, 1868–1911.

Cimarron (N.Mex.) News, 1874.

Cimarron (N.Mex.) News and Press, 1875–78.
Colorado Chieftain, 1877.
Columbus (Iowa) Safeguard, 1896–98.
Denver Tribune, 1885.
Elizabeth City (N.Mex.) Railway, Press, and Telegraph, 1874.
Las Vegas (N.Mex.) Gazette, 1883–86.
Las Vegas (N.Mex.) Daily Optic, 1884–1905.
Las Vegas (N.Mex.) Optic, 1910–11.
New York Sun, 1875–76.
New York Times, 1887.
Raton (N.Mex.) Comet, 1882–86.
Raton (N.Mex.) Guard, 1882.
Raton (N.Mex.) Daily Range, 1887.
Raton (N.Mex.) Range, 1955.
Raton (N.Mex.) Weekly Independent, 1886.
Santa Fe Daily New Mexican, 1874–90.
Santa Fe New Mexican, 1890–1927.
Santa Fe Sentinel, 1879.
Santa Fe Weekly New Mexican, 1874–78.
St. Louis (Mo.) Daily Globe, 1886.
Trinidad (Colo.) Daily News, 1878.
University (of Iowa) Reporter, 1870.
Wapello (Iowa) Republican, 1898.
Washington Post, 1928.

Records of Legal Proceedings

Cimarron Valley Land Company v. Mary Gallagher et al. Case No. 3106, Eighth Judicial District of New Mexico, New Mexico State Records Center and Archives, Santa Fe.
Cunningham v. Springer 13 N.Mex. 259 (1905).
Cunningham v. Springer 204 U.S. 647 (1907).
In re Catron 8 N.Mex. 253 (1895).
Interstate Land Company v. Maxwell Land Grant Company et al. 139 U.S. 569 (1891).
Maxwell Land Grant Company v. Dawson 151 U.S. 586 (1894).
Maxwell Land Grant Company v. John B. Dawson 7 N.Mex. 133 (1893).
Springer Land Association v. Ford 168 U.S. 513 (1897).
The State v. Lawrence, 38 Iowa Reports 51 (1873).
Tameling v. United States Freehold and Emigration Company 93 U.S. 644 (1877).
Thompson v. Maxwell Land Grant and Railway Company 168 U.S. 451 (1897).
United States v. Maxwell Land Grant Co. 5 N.Mex. 297 (1889).

United States v. Maxwell Land-Grant Co. et al. 21 Federal Reporter 19 (1884).

United States v. Maxwell Land Grant Co. et al. 26 Federal Reporter 118 (1886).

United States v. Maxwell Land-Grant Co. et al. 121 U.S. 325 (1887). Transcript of Record, Case File 12100, National Archives and Records Administration, RG 267.

United States v. Maxwell Land-Grant Co. et al. 122 U.S. 365 (1887).

Index

Page numbers in *italic* typeface refer to illustrations.

ISBN 1-58544-464-2
53495
9 781585 444649